Scottish Wild Flowers

Flowers

Mini-Guide

Michael Scott

BIRLINN

First published in 2012 by
Birlinn Limited
West Newington House, 10 Newington Road
Edinburgh EH9 1QS
www.birlinn.co.uk

Illustrations from *Collins New Generation Guide to the Wild
Flowers of Britain and Northern Europe*
Additional artwork by Valerie Price and Sue Scott
Photographs by the author unless otherwise stated

ISBN-13: 978 1 84158 954 1

British Library Cataloguing-in-Publication Data
A catalogue record for this book is available from the British
Library

For Maisie and Archie – the next generation

*Printed and bound in India
by Associated Agencies Ltd
Witney, Oxford*

Contents

About This Book

This Mini Guide is a stripped-down version of Scottish Wild Flowers, also published by Birlinn Ltd. While that book fits comfortably in the pocket, this volume is suitable for a handbag, the car glove box or the corner of a rucksac. The main focus is on plants of the low-lying areas of Scotland, but the final section looks at the commoner plants that visitors may encounter on the moorlands and uplands that are such a feature of Scotland.

This *Mini Guide* covers nearly 300 species. Almost every lowland plant in the larger guide is included, but with less space per species. That has meant losing information on some aspects of cultural history that make the study of Scotland's wild flowers so fascinating. There is no space for Gaelic names, or for a gazetteer of places to visit, which feature in the larger guide. Most importantly, the parent volume includes information on many additional species, similar to those illustrated, but they have had to be dropped from this volume.

Scots Primrose grows in clifftop grassland in Caithness, Sutherland and Orkney and nowhere else in the world.

However, this *Mini Guide* has the huge benefit of being quick to use and easy to carry. By referring to the text and the artwork, readers should be able to identify, with reasonable confidence, the commonest Scottish flowering plants. The main identification features of each species are shown in *italics*.

Why do we need a Scottish guide? Well, Scotland has some special plants, like Scots Primrose (p. 116) that grows nowhere else in the world. Other species like Scots Lovage (p. 123) reach virtually their southernmost sites in Scotland. Visit many of the fine plantation woods around the Lowlands in spring with a traditional flower guide, and you will be instantly baffled. These woods have many non-native species that are largely absent elsewhere in the UK, but they are covered fully in this guide.

A Brief Guide to Terminology

Although technical terminology is kept to a minimum, it is impossible to describe a flower without naming its parts, as illustrated in the Creeping Buttercup (right). It is helpful to consider the flower as a series of concentric rings of 'leaves' which become more specialised towards the centre. Thus, the outer **sepals**, which protect the flower in bud, are generally like a leaf in shape and colour. Inside these, the ring of **petals** retain a leaf-like shape, but are usually brightly coloured to attract pollinating insects. Plants whose pollen is dispersed by the wind therefore need only inconspicuous petals. In some flowers, the sepals and/or petals are united into a **tube.**

The club-shaped **stamens** within the petals carry pollen (male sex-cells). The innermost 'leaves' form one or more **ovaries**, enclosing the female sex-cells. The ovary is topped by a pollen-receptor, the **stigma**, which may be held on a stalk called the **style**. After fertilisation, the ovules develop into seeds and the ovary swells into a **fruit**. This need not be fleshy, but can be a pouch-like **capsule** or a **pod** like a pea-pod.

Beneath the **inflorescence** (the branching or clustered grouping of flowers), some plants have a further set of leaves called **bracts**. These may or may not be the same shape as other stem leaves.

Other unavoidable terminology is the distinction between **shrubs**, which have wood in their stems, and **herbs** which lack woody stems. Finally three terms are used to describe different plant lifecycles: **annuals** flower briefly then set seed and die, with only their seeds surviving to grow in the next flowering season; **perennials** survive through several growing seasons, although not necessarily flowering in each, and **biennials** complete this lifecycle in two years.

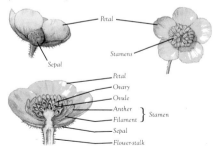

How to Use This Book

The book is divided into six colour-coded sections, arranged according to the main Scottish lowland habitats. Within each section, the order of species follows the most recent scientific arrangement. The common name given is generally widely used in Britain, although, in a few cases where Scots names are in wide usage, these are preferred. Scientific names are based on the standard British work (*New Flora of the British Isles*; Stace, 2010).

For each species, the flowering season (**Fl**) is given (where 1 = January and 12 = December), and the typical height (**Ht**) of the plant, although where a plant climbs or scrambles, this may refer to the length of the stem instead.

The geographical range of each species refers to Scotland only, and is based on maps in the *New Atlas of the British Flora* (2002). Abbreviations have been minimised in the descriptions, other than the use of N, E, S and W for directions, and C for central. N Isles refers to Orkney and Shetland, W Isles to the Outer Hebrides and W islands to all the islands on the western seaboard of Scotland. Scotland includes the Isles, unless otherwise stated. Other areas use traditional county names.

Responsible Access

The Land Reform (Scotland) Act 2003 gives everyone a right of access on foot to almost all land in Scotland, provided that right is exercised with responsibility. The exceptions are houses and their immediate surroundings, land in which crops are growing (although you can walk round field margins), sports fields, golf courses and sites for which access is charged. With that right comes the responsibility to respect people's privacy, to recognise and assist the legitimate activities of farmers and landowners, and to care for the environment.

The Scottish Outdoor Access Code (*www. outdooraccess-scotland.com*) gives full details of these rights and responsibilities.

Heather is very much part of the Highland landscape, turning glorious purple in late summer.

A Botanical Code of Conduct

- Do not dig up any wild plants.

- Always leave wild flowers for others to enjoy.

- To identify a flower, take this guide to the plant, not *vice versa*.

- If a sample is essential to identify a plant, take the smallest adequate piece **only** if the plant is common.

- Take great care when studying or photographing one plant not to accidentally trample another.

- If a rare plant is discovered, avoid exposing it to unwelcome attention by making a path or flattening vegetation around it, and do not reveal its locality to anyone who will not respect that confidence.

- Always consider the needs of other wildlife and, in particular, avoid disturbance to breeding birds.

PLANTS AROUND PEOPLE

The plants that live closest to people in Scotland –
in cities and wasteground, fields and roadsides – are
the least markedly Scottish in this book. Plants have
been carried so widely by human activities that
most of these 'domesticated' plants will be familiar
to anyone from mainland Europe and several even
to American visitors.

Indeed, many of the most conspicuous flowers
of wasteground in Scottish cities are not native
here. For example, the Butterfly-bush, which
can colonise the most inhospitable city gap sites,
is a native of China which escaped from gardens
early this century. Similarly, Wallflower, which
blooms abundantly over Edinburgh Castle Rock in
May and perches on several other Scottish castles,
originated in the western Mediterranean.

Perhaps the most remarkable story is that of
Oxford Ragwort, now one of the commonest
plants on untended walls, cracked pavements and
wasteground in Edinburgh and Glasgow. A native

A dense stand of Rosebay Willowherb

of Sicily and southern Italy, it was brought to the botanic garden in Oxford in the 18th century. By 1794, it had escaped onto walls nearby, and from there it began to spread slowly. The clinker of railway lines provided a perfect habitat, and trains helped carry its parachuted seeds. It reached Edinburgh around 1948, Dundee by 1967 and as far north as Shetland in 1986.

Not every city plant is an introduction from exotic parts. Rosebay Willowherb is perhaps

PLANTS AROUND PEOPLE

the most abundant plant of disturbed ground in Scottish cities. It is native, but 80 years ago it was only found in damp woods and rocky places. The burnt-out bomb sites of the Second World War provided exactly the right sort of damp, nitrate-rich soil for it to establish, and – with each plant producing 80,000 seeds which float in the wind like a snowstorm – it soon spread more widely.

Outside cities, any drive around Scotland in spring or summer is enlivened by the roadside flowers. Many roadside verges are linear, left-over fragments of wild vegetation that once flourished in the area – woodland or heathland for example. Lime-loving plants such as Cowslip, Crosswort and Bloody Cranesbill add colour to some coastal roads in east Scotland, while quiet verges in the far north-west can provide a display of orchids to rival any garden.

Increasingly, wildflower seed is being planted into the verges of new roads to soften their impact on the landscape. Seeds of Red or White Clover are often sown in newly-laid soil because of the enriching effects of their nitrogen-fixing roots. Other wild flowers are included for their beauty and in recognition of the value of verges as wildflower refuges. Some species appear of their own accord. The seeds of Common Poppy, for example, can live in the soil for 20 years, ready to spring up after road-building machinery has passed.

Beyond the verges, arable fields seem bare in comparison. Herbicides and cleaner seed-corn have eliminated the colourful (but troublesome) displays of wild flowers that once graced field margins. Occasionally, a field may miss its spring spraying, and a glorious sheet of red Poppies or yellow Corn Marigold may briefly brighten the countryside, although nowadays a field of yellow is more likely to be cultivated Rape.

In north-west Scotland, however, traditional, low-intensity pastoral agriculture still survives,

It is rare today to see such a massed display of Corn Marigold, a victim of herbicides and cleaner seed-corn

PLANTS AROUND PEOPLE

A rich display of colourful weeds around a crofter's potato field in South Uist are a reminder of fields before the advent of herbicides

supported by the uniquely Scottish system of crofting in which a part-time farmer has heritable and legally-protected tenant's rights to a small-holding. Here, rather more of the traditional field-edge plants are left to flower.

The best displays are found on the exposed western seaboard of the Outer Hebrides and Sutherland. The prevailing winds here carry white shell sand from the beaches inland to cover the underlying peat. This produces a thin soil which the crofters fertilise with seaweed and plough shallowly to grow barley or potatoes, with long periods left fallow. The result is the machair, the flat, fertile fringe of the crofting lands which still supports a colourful abundance of flowers in summer.

Common Poppy
Papaver rhoeas
POPPY FAMILY

Fl: 6-9 Ht: 20-60 cm

This attractive annual is widespread in disturbed ground and waste places in S Scotland, although herbicides have reduced its abundance. It is less common in the N and W (where other species grow). Its showy *flowers are 5-10 cm across* with 4 *scarlet petals* that are often *dark-blotched* at the base. Its pepperpot-like seed capsule is *roundish to egg-shaped*.

Welsh Poppy
Meconopsis cambrica
POPPY FAMILY

Fl: 6-9 Ht: 30-60 cm

This *hairless, yellow-flowered* poppy, native to Wales, SW England, Spain and France, is widely grown in gardens. Escaped plants are sometimes found in sheltered sites near habitation, mostly in the Borders and lowlands. Its leaves are *divided into toothed, pointed oval lobes*, and its *egg-shaped* fruiting capsule splits into *short teeth at the tip*.

Common Fumitory
Fumaria officinalis
POPPY FAMILY
Fl: 5-10 Ht: climbing to 1 m

This is the commonest of several very similar Fumitories which scramble over other plants in field margins on light soils. They have *delicate, feathery, greyish-green leaves*, and spike-like heads of about 20 *tubular pink flowers*, 7-8 mm long, *tipped with deeper crimson*. This species is most frequent in E Scotland, less common in the SW and scattered in the Isles.

Broom
Cytisus scoparius
PEA FAMILY
Fl: 5-6 Ht: 60-200 cm

Broom differs from Whin (p.109) in having *no spines* but having *green, strongly angled twigs* with *stalked, 3-lobed leaves*, which are shed in winter. Its *golden-yellow flowers*, to 2 cm long, grow in the angles of leaves, and ripen to explosive *black pods*. It grows in sandy heaths, woods and wasteground with poor soils around Scotland, but is planted in the Isles.

BUTTERCUPS

Most buttercups (*Ranunculus* species) are perennials with leaves that are deeply cut into toothed lobes and often hairy. The flowers typically have 5 golden-yellow, overlapping petals (a) and 3–5 green sepals (b), ripening into a tight cluster of single-seeded, egg-shaped fruits held in a roundish head (c). They flower in spring to summer.

Meadow Buttercup
Ranunculus acris
BUTTERCUP FAMILY
Fl: 5-8 Ht: 15-100 cm

The commonest buttercup, this grows in damp meadows and pastures, roadside verges, ditches and open woodland throughout Scotland. Its flowers are *15–25 mm across*, with 5 green *sepals held upright against the flower*. Its *flower-stalks are unfurrowed* and its *softly-hairy leaves* are divided into *3–7 narrow, wedge-shaped lobes*, which are *roundish in outline*.

Creeping Buttercup
Ranunculus repens
BUTTERCUP FAMILY
Fl: 5-8 Ht: 15-60 cm

Even commoner than Meadow Buttercup in heavy, nutrient-rich soils, this species grows in wet grassland, woods and ditch-sides around the country. It spreads by *creeping, surface runners* which *root at intervals*. Its flowers are larger (*20–30 mm in diameter*), on *furrowed, hairy stalks*, usually with 5 or 6 petals and *upright or spreading sepals* (see box opposite).

Bulbous Buttercup
Ranunculus bulbosus
BUTTERCUP FAMILY
Fl: 4-6 Ht: 15-40 cm

The least common of the 3 field buttercups, Bulbous Buttercup grows scattered throughout the lowlands, particularly near coasts. It prefers drier and more lime-rich pastures, but is intolerant of trampling. It is distinguished by its *furrowed, hairy flower-stalks*, *down-turned sepals*, leaves with a *long-stalked middle lobe*, and *swollen, bulbous stem base*.

Bramble
Rubus fruticosus agg.
ROSE FAMILY

Fl: 5-9; Ht: variable to 1.2 m

Specialists recognise 300 'micro-species' of the common wayside 'Blackberry'. Generally, they have *arching, prickly stems, toothed, 3-lobed leaves*, 5-petalled white or pink flowers, and *glossy black, raspberry-like fruits*. They grow in bushy places, woods and roadsides in most of Scotland, although more scattered in the Highlands and W Isles, and introduced to the N Isles.

Dog Rose
Rosa canina
ROSE FAMILY

Fl: 6-7 Ht: 1-3 m

The commonest of several wild roses in woods, hedges and scrub around most of Scotland, this bushy shrub has *prickly, arching stems*. Its leaves have *1–3 paired leaflets* and a single end leaflet. Its pink or white flowers are 3–5 cm across, with 5, often *notched, petals*. These ripen into *rose hips*, in which the swollen flower base encloses the true fruits ('pips') inside.

PLANTS AROUND PEOPLE

Common Nettle
Urtica dioica
NETTLE FAMILY
Fl: 6-8 Ht: 30-150 cm

Virtually universal in Scotland, this perennial forms dense patches in hedgebanks, woods, grassy places, and disturbed ground, especially near buildings. Its leaves are *coarsely-toothed* and usually covered in *stinging hairs* which break when grasped, injecting formic acid. *Tiny, wind-pollinated flowers hang in tassels* in the angles of the upper leaves.

Sun Spurge
Euphorbia helioscopia
SPURGE FAMILY
Fl: 5-10 Ht: 10-50 cm

The 'flowers' of Spurges are actually a cluster of *tiny, petal-less flowers*, enclosed within a greenish bract which has *conspicuous lobes or glands*. Sun Spurge is declining in cultivated, lowland areas around Scotland, and commonest now on W Isles machair. It has *blunt-ended, finely-toothed leaves*, which *taper to a narrow base*, and *round-lobed bracts*.

Field Pansy
Viola arvensis
VIOLET FAMILY

Fl: 5-10 Ht: stems 15-45 cm

This typical pansy (p.76) is distinguished from Wild Pansy (p. 77) by its smaller flowers *(up to 20 mm vertically)* and creamy-white or violet-tinted petals, *shorter than the pointed sepals.* The spur behind the flower is *no longer than the backward-pointed flaps* at the sepal base. It grows in cultivated and waste ground in E Scotland, W coastal regions and the Isles.

Broad-leaved Willowherb
Epilobium montanum
WILLOWHERB FAMILY

Fl: 6-8 Ht: 5-60 cm

Willowherbs (p. 165) have *4 rose-coloured petal-lobes* on top of a *long flower tube*, which extends after fertilisation into a *narrow, cylindrical fruiting capsule.* This is the commonest Scottish species in hedgerows, woods, rocks and gardens throughout the country. It is distinguished by *hairless, broadly egg-shaped, stalked leaves* and by a *cross-shaped stigma.*

Rosebay Willowherb
Chamerion angustifolium
WILLOWHERB FAMILY
Fl: 7-9 Ht: 30-120 cm

This handsome perennial (see p. 12) is commonest in wasteground and urban gap sites throughout Scotland, as well as damp woods and rocky places. Its creeping roots produce *dense patches of slightly hairy stems*, with many *lance-shaped leaves*. Its *deep-pink flowers*, born in *dense spikes*, have four petals, but with the *upper two petals broader than the lower two*.

Common Mallow
Malva sylvestris
MALLOW FAMILY
Fl: 6-9 Ht: 45-90 cm

This robust perennial has a *sparsely hairy stem*, and *roundish, gently-lobed leaves*. In late summer, its clustered *pinkish-purple flowers* are up to 4 cm across, with *purple-veined petals* that are *deeply notched at the tip*. It grows on roadsides and waste places in SE Scotland and the Moray Firth area, with a few scattered W coast sites.

Wood Cranesbill
Geranium sylvaticum
CRANESBILL FAMILY
Fl: 6-7 Ht: 50-80 cm

Scattered in upland meadows, hedgerows and woods in the Borders and C Highlands, this showy plant has *reddish-mauve flowers,* up to 3 cm in diameter, ripening to *erect-growing fruits.* Its long-stalked leaves usually have 7, *toothed or shallowly cut* lobes. In lowland and S Scotland, it is replaced by Meadow Cranesbill (*G. pratense*) with *violet-blue* flowers.

Dovesfoot Cranesbill
Geranium molle
CRANESBILL FAMILY
Fl: 4-10 Ht: 10-40 cm

This annual has *densely hairy stems* and *rounded, softly hairy, 5- to 9-lobed* leaves. Its small flowers, which grow in pairs on longish stalks, are *less than 1 cm across,* with *deeply-notched, rosy-purple petals* and *hairy sepals.* It grows in dry grassland, cultivated land, sand dunes and waste places all round Scotland, but is uncommon in the C and NW Highlands.

PLANTS AROUND PEOPLE

Herb-Robert
Geranium robertianum
CRANESBILL FAMILY
Fl: 5-9 Ht: 10-50 cm

This straggly annual of open woodland and hedgerows has *reddish stems* and *deeply-lobed leaves* with only *scattered hairs*. Its *flowers* are often paired, *up to 2 cm across*, with narrow *pink* petals, *not notched* at the tip. Their anthers are *orange or purple*. It is found around Scotland, but more locally in the N, uncommon on the W Isles, and introduced to the N Isles.

Wallflower
Erysimum cheiri
CABBAGE FAMILY
Fl: 4-6 Ht: 20-60 cm

This familiar garden plant from Greece, with *bright yellow flowers*, has been established since medieval times on castle walls and old stonework in the Borders, E and SW. Typical of the Cabbage Family (see p.26), it has *4 petals*, but no wild relative has *such large and showy flowers*, with *petals up to 2 cm long*, on stems with crowded, *hairy, narrowly-oblong leaves*.

CABBAGE FAMILY

Members of this family, like Turnip (right), are called crucifers from the typical cross shape of their flowers, which normally have 4 not united petals, 4 sepals and 4 or 6 stamens. They ripen into fruit capsules which are cylindrical or oval in shape. Their leaves are usually lobed or toothed. Flower colour, leaf and capsule shape are the main identification characters.

Winter-cress
Barbarea vulgaris
CABBAGE FAMILY
Fl: 5-8 Ht: 30-90 cm

Winter-cress grows in fields, waste ground, riverbanks and roadside verges around S and E Scotland. The *large, oval terminal lobe* of its divided, glossy-green leaves distinguish it from all other yellow-flowered crucifers. The short sepals are *held tightly against* the petals of the small flowers, which are *7–9mm across*. The *fruit pod* is a flattened or *4-angled* cylinder.

Shepherd's Purse
Capsella bursa-pastoris
CABBAGE FAMILY
Fl: 1-12 Ht: 3-40 cm

This annual or biennial crucifer flowers almost throughout the year in cultivated land, roadsides, wasteground and sand dunes. Its *leaves*, which form a *neat rosette* on the ground or *clasp the stem*, are *spear-shaped* in outline, but vary from deeply-toothed to almost undivided. The *tiny, white flowers, 2–3 mm in diameter*, develop into *heart-shaped pods*.

Wavy Bitter-cress
Cardamine flexuosa
CABBAGE FAMILY
Fl: 4-9 Ht: 10-50 cm

Wavy Bitter-cress has zigzag stems with a *few-leaved basal rosette* and many *roundly-lobed stem leaves*. Its flowers, c. *4 mm across with 6 stamens*, develop into a *long, narrow pod*. It grows in moist, shady places through most of Scotland, but is a garden outcast in Shetland. Hairy Bitter-cress (*C. hirsuta*), with *few stem leaves* and *4 stamens*, has a similar range.

Smith's Pepperwort
Lepidium heterophyllum
CABBAGE FAMILY
Fl: 5-8 Ht: 15-45 cm

This hairy perennial is found in cultivated places and waysides, scattered throughout Scotland, except for the N and Isles. Its *narrow, toothed leaves* cluster round and *clasp the stem*, which is topped in summer by *long, crowded heads* of *small, white flowers*. These develop into *oval, 2-seeded fruits* with a *short style in a notch at the tip*, and *broad wings*.

Rape
Brassica napus
CABBAGE FAMILY
Fl: 4-8 Ht: up to 1 m

At least 36,000 hectares of Rape was grown in Scotland in 2010. Because its *beaked, cylindrical fruits* need to mature before harvest, some seeds inevitably are spread, and Rape is now one of the commonest yellow crucifers in roadsides and wasteground around the lowlands. Its *pale-yellow flowers* sit *below the developing buds*. Its leaves are *bluey-green*.

Charlock

Sinapsis arvensis

CABBAGE FAMILY

Fl: 5-7 Ht: 30-80 cm

Charlock grows in arable and disturbed ground throughout agricultural Scotland. Its much-branched *stems* are covered in *down-turned hairs*. Its *spear-shaped leaves* are often purply, the *upper ones without stalks*. The sepals of its bright-yellow flowers *spread outwards*. The bristly *fruiting pod* ends in a *straight beak, about half as long as the seeded portion.*

White Mustard

Sinapsis alba

CABBAGE FAMILY

Fl: 6-8 Ht: 30-80 cm

The mustard of mustard-and-cress, this Mediterranean annual is grown as a forage crop in lowland areas, and is sometimes accidentally introduced with grain or bird seed, mainly in E and NE Scotland. It resembles Charlock but has *stalked* leaves with *many, irregular lobes*, and its hairy pod ends in a *flattened, curved beak, at least as long as the seeded portion.*

Wild Radish
Raphanus raphinastrum

CABBAGE FAMILY

Fl: 5-9 Ht: 20-60 cm

This annual of cultivated land, wastexground and roadsides is found throughout the lowlands and N Isles, but less commonly in the W. Its leaves are *deeply-lobed*, with a *large end lobe and side lobes rapidly shrinking towards the leaf base*. Its flowers resemble Charlock (p. 29), but are *yellow, white or lilac* and its cylindrical pods are *constricted between the seeds*.

Garlic Mustard
Alliaria petiolata

CABBAGE FAMILY

Fl: 4-6 Ht: 20-120 cm

Less common in Scotland than England, this handsome plant grows in roadsides, hedgerows and woodland edges, but is absent in the NW and Isles. Its *glossy, pale-green, heart-shaped* leaves *smell strongly of garlic* when crushed. Its stems are topped by a cluster of *white, 4-petalled flowers*, about 6 mm across, which develop into a *long, narrow, up-curved fruit pod*.

Hedge Mustard
Sisymbrium officinale
CABBAGE FAMILY
Fl: 6-7 Ht: 30-90 cm

This annual or overwintering weed grows in hedgebanks, fields, roadsides and wasteground throughout Scotland, except for the C and NW Highlands. Its stem stands stiffly erect, *with wiry branches almost at right-angles*. Its tiny flowers (*about 3 mm across*) have *pale-yellow petals*, and develop into long, rather hairy pods which are *pressed tightly against the stem*.

Redshank
Persicaria maculosa
DOCK FAMILY
Fl: 6-10 Ht: 25-75 cm

Redshank is an abundant annual weed of arable fields, wasteground, waysides and marshland around Scotland. Its reddish, hairless stem is swollen and encircled by a *membranous sheath, topped with bristles,* at the base of each lance-shaped, often *dark-blotched* leaf. In summer, the stem is topped by a dense spike of *pink flowers*.

Knotgrass
Polygonum aviculare
DOCK FAMILY
Fl: 7-11 Ht: creeping to 2 m

Another abundant annual of disturbed ground, arable fields and gardens, Knotgrass has a similar distribution to Redshank (p.31). It is much-branched and often spreads low over the ground, with *oval leaves* which are larger on the main stem than on the flowering branches. It has small, *knot-like clusters of tiny pink or white flowers* in the angles of upper leaves.

Japanese Knotweed
Fallopia japonica
DOCK FAMILY
Fl: 8-10 Ht: up to 2 m

Introduced to gardens from Japan and escaping by 1825, this aggressive alien is widespread in S and E Scotland, and more localised elsewhere including the Isles. Its *wavy, reddish stems* die back each year but regrow rapidly each spring. Its *oval leaves* are 6–12 cm long with *a long tip* and *sharply cut-off base*. Its *dingy-white flowers* grow in tassels between the leaves.

PLANTS AROUND PEOPLE

Giant Knotweed
Fallopia sachalinensis
DOCK FAMILY
Fl: 8-9 Ht: rarely to 4 m

Even taller than Japanese Knotweed (opposite), this species was grown in fewer gardens. As a result, it escaped less into the wild, with just a few scattered, mainly coastal sites, including on the W Isles. Its *leaves* are *longer* (15–30 cm), *pointed, but less drawn-out at the tip* and often *heart-shaped* at the base. It has *shorter, denser* inflorescences of *greenish flowers*.

Black Bindweed
Fallopia convolvulus
DOCK FAMILY
Fl: 7-10 Ht: scrambling 30-120 cm

The *stalked, heart-shaped leaves* and *twining stems* of this scrambling annual resemble true Bindweeds (p. 42), but its flowers are *small and inconspicuous*, with 5 *greenish-white* petals. They ripen into *dull black, triangular fruits*. It is a frequent weed of arable land, wasteground and gardens, mostly in E Scotland, but rarer in the N and W, including the islands.

DOCK FAMILY

Known as dockens in Scotland, members of this family are not easy to separate at first because their spikes of small, greenish, wind-pollinated flowers are so reduced. The best distinguishing features are the shape of their undivided leaves and the inner 'sepals', which enlarge into hard valves around the reddish, 3-sided fruits and often bear prominent warts.

Curled Dock

Rumex crispus

DOCK FAMILY

Fl: 6-10 Ht: 50-100 cm

Curled Dock is found in waste places, cultivated land, pastures and wet ground throughout Scotland, including the Isles, except for some parts of the C and NW Highlands. It has *wavy-edged, pointed, lance-shaped leaves* to 25 cm long, and *rather leafy flowering spikes*. Its *heart-shaped fruit valves* are *toothless* and 1, 2 or all 3 of them have a *smooth, oblong wart*.

Broad-leaved Dock
Rumex obtusifolius
DOCK FAMILY
Fl: 6-10 Ht: 50-100 cm

This species has a similar distribution to Curled Dock (opposite) but is less abundant in pastures, preferring open habitats like field margins, ditches, roadsides and waste ground. It has *broader, more egg-shaped, blunt-tipped leaves*, to 15 cm across, which are often *heart-shaped at the base*. Its triangular fruit valves are *strongly-toothed with a round wart*.

Clustered Dock
Rumex conglomeratus
DOCK FAMILY
Fl: 7-10 Ht: 30-60 cm

Slighter than the previous species, this dock has slender, wavy stems with *widely-spreading branches* and *narrowly oval leaves*. Its flowers and fruits are borne in *knot-like clusters* on branches which are *leafy almost to the tip*. Its oval *fruit valves* have *large, swollen warts*. It grows mainly in marshy meadows, ditches and stream-sides in C and E Scotland.

Thyme-leaved Sandwort

Arenaria serpyllifolia

PINK FAMILY

Fl: 6-8 Ht: 3-25 cm

This slender, much-branched annual grows in open ground and gardens scattered over most of Scotland, but is rarer in the N and Isles. Its *greyish-green, pointed, oval leaves are covered in rough hairs,* like its stems. It has a branched inflorescence of many small flowers (*to 8 mm across*) with *undivided white petals, shorter than the sepals,* ripening to *flask-shaped fruits.*

Common Chickweed

Stellaria media

PINK FAMILY

Fl: 3-12 Ht: 5-40cm

This annual of cultivated ground, roadsides and waste places around Scotland is recognised by the *single line of hairs* down its straggly stems between the opposite pairs of *egg-shaped, hairless leaves,* the lower of which are long-stalked and the upper stalkless. Its small flowers have *white petals a little shorter than the sepals* and 3–8 stamens with *purplish anthers.*

Common Mouse-ear
Cerastium fontanum

PINK FAMILY

Fl: 4-9 Ht: creeping to 40 cm

This is the commonest of several similar annuals or short-lived perennials, widespread in cultivated ground and grassland around Scotland. All have weak, sprawling stems, *oval leaves with whitish hairs* (like mouse ears), and flowers with *5 sepals* and *5 notched, white petals*. This species has especially *deeply notched petals, just a little longer than its sepals.*

Procumbent Pearlwort
Sagina procumbens

PINK FAMILY

Fl: 5-9 Ht: spreading to 20cm

Because this *mat-forming perennial* survives trampling and likes nutrient-rich soils, it grows abundantly in paths, lawns, arable fields and wet, stony ground all round Scotland. Its *rooting branches, with moss-like leaves*, creep along the ground. Its tiny flowers, in summer, have *4 (rarely 5) green sepals* and either *minute white petals or none at all.*

Corn Spurrey

Spergula arvensis

PINK FAMILY

Fl: 6-9 Ht: 8-40 cm

This often troublesome weed of sandy soil in fields, tracks, gardens and wasteground is found in all but the Scottish uplands. It has a *straggling stem*, often covered in *sticky hairs*, and *narrow, fleshy, rather greyish leaves* in *tufts* up the stem. Its flowers have 5 rather narrow, undivided, white petals, which are *slightly longer than the sepals* between them.

Red Campion

Silene dioica

PINK FAMILY

Fl: 5-7 Ht: 30-90 cm

Red Campion enlivens hedgerows, wood margins and sea-cliffs over most of Scotland, but is uncommon in the NW and W Isles. It has *robust, hairy* stems with *oval to oblong leaves*. Its *deep pink*, tubular flowers are about 2 cm across, with *hairy sepal-tubes*, and spreading *petal-lobes* that are *deeply notched at the tip*. Shetland plants are sturdier and deeper red.

PLANTS AROUND PEOPLE

White Campion

Silene latifolia

PINK FAMILY

Fl: 5-9 Ht: 30-100 cm

Similar to Red Campion with which it often hybridises, but with *white, evening-scented flowers*, this species is more restricted to the Lowlands and E, with few sites in the W. It grows in fields, wasteground, roadsides and grassy banks. It has a *sticky-hairy stem* and a swollen fruiting capsule with 10 *erect teeth* at the tip (down-turned in Red Campion).

Fat-Hen

Chenopodium album

GOOSEFOOT FAMILY

Fl: 7-10 Ht: up to 1.5 m

In the same family as Spinach and Beet, Fat-Hen is an annual with tiny flowers, arranged in narrow spikes on *reddish stems*. Its rather fleshy *leaves vary from spear- to egg-shaped* with at least the lower ones toothed, and often have a *white, mealy covering*. It grows in waste places and cultivated land, scattered through the lowland areas of Scotland.

Field Madder

Sherardia arvensis

BEDSTRAW FAMILY

Fl: 5-10 Ht: 5-40 cm

This delicate, trailing annual is found in dry arable fields, open grassland and hedgebanks in the lowlands of E Scotland and near coasts in the W. It has whorls of 4–6, *oval, pointed, bristle-edged leaves* around a rather weak, squarish stem. This is *prickly at its angles* and tipped with a small, clustered head of 4–6 *pale-lilac flowers* with 4 spreading petal lobes.

Hedge Bedstraw

Galium album

BEDSTRAW FAMILY

Fl: 6-7 Ht: 25-120 cm

Bedstraws typically have whorls of 4 or more narrow leaves around square stems, and small tubular flowers with 4 joined petals. The *scrambling* stem of this species is *swollen beneath each whorl of 6–8 prickle-edged leaves.* Its clustered white flowers are *rarely more than 4 mm across.* It grows in hedgebanks, grasslands, scrub and wasteground in S and E Scotland.

Cleavers
Galium aparine
BEDSTRAW FAMILY
Fl: 6-8 Ht: scrambling 15-120 cm

Cleavers or Goosegrass – known as 'Sticky Willie' in Scotland – grows abundantly in hedges and wasteground throughout the country. Its *weak, scrambling* stems break readily, and like its leaves and peppercorn-sized fruits, are covered in *hooked bristles*. This helps to spread the plant tangled in animal hair. Its flowers are *pale green* and *no more than 2 mm across*.

Bugloss
Anchusa arvensis
BORAGE FAMILY
Fl: 6-9 Ht: 15-50 cm

This *bristly-haired* annual grows in sandy fields in E Scotland and the N Isles, more rarely in the W and most commonly on W Isles machair. It has *narrow, tongue-shaped leaves*, and flowers with 5 spreading, *bright blue petal lobes*, united at their base into a tube which is *kinked in the middle* with *conspicuous white scales* at its throat. (cf Green Alkanet, p. 140).

Field Forget-me-not
Myosotis arvensis
BORAGE FAMILY
Fl: 4-9 Ht: 15-30cm

Forget-me-nots have tubular blue flowers with spreading petal-lobes, in inflorescences curled like scorpions' tails. This species is common in cultivated land, wasteground, woods and roadsides throughout Scotland. It has downy, lance-shaped leaves and flowers *less than 4 mm in diameter,* on *stalks longer than the sepal-tubes* which are covered in *hooked hairs.*

Field Bindweed
Convolvulus arvensis
BINDWEED FAMILY
Fl: 6-9 Ht: scrambles to 75cm

True bindweeds typically have twining stems that scramble over vegetation or fences, and showy, funnel-shaped flowers. This relatively delicate plant has *arrow-shaped* leaves and small, white or pink flowers, *2–3 cm across,* with *small bracts* held *below the true sepals.* It is a troublesome weed of cultivated land, wasteground and hedgebanks in E and C Scotland.

Hedge Bindweed
Calystegia sepium
BINDWEED FAMILY

Fl: 7-9 Ht: climbing to 2 m

This is the most widespread Scottish Bindweed, growing in hedges and bushy places in the lowlands, but nowhere commonly. It is more robust than Field Bindweed (opposite) with weak stems and large *heart-shaped leaves*. Its white or pinkish flowers are *3–4 cm across*. Below them are *2 sepal-like bracts, about 1.5 cm wide*, enclosing the true sepals beneath.

Bittersweet
Solanum dulcamara
NIGHTSHADE FAMILY

Fl: 6-9 Ht: scrambling to 3 m

This shrubby plant is found only in S Scotland and around the Moray Firth, and as an introduction on Skye and Orkney. Its weak, downy stems scramble through hedgerows and scrub. It has *oval leaves*, and *flowers* with a *yellow central column* and *5 purple petal-lobes which bend backwards* as the flowers mature. These ripen from green to red berries.

Foxglove
Digitalis purpurea
FIGWORT FAMILY
Fl: 6-9 Ht: 15-150 cm

This familiar flower grows commonly in woods, heaths and rocky places throughout Scotland, although only a garden outcast on Shetland. It is biennial, producing *downy, tongue-shaped leaves* in its first year and a flowering stem in its second summer, with 20–80 *thimble-like flowers* that are *deep pink outside and creamy, with purple spots, inside.*

Ivy-leaved Toadflax
Cymbalaria muralis
SPEEDWELL FAMILY
Fl: 5-9 Ht: trailing to 60cm

This attractive plant of sunny old walls has *rather fleshy, often purplish, ivy-shaped leaves* and *lilac-and-yellow snapdragon-like flowers.* Its delicate *purple stems* trail over the wall, and root in the mortar. Introduced from Italy and the Alps since 1640, it is widespread in S and C Scotland, more scattered in the NE and W coast, and generally absent from the Isles.

Common Toadflax
Linaria vulgaris
SPEEDWELL FAMILY
Fl: 6-10 Ht: 30-80 cm

Although nowhere frequent, this handsome plant brightens roadsides, wasteground and grassland around the lowlands and E Scotland. It spreads by creeping runners, occasionally forming dense patches. It has *narrow, lance-shaped, greyish leaves* and long spikes of bee-pollinated *snapdragon-like yellow flowers,* with a *3-lobed lower lip and an orange 'mouth'*.

Butterfly-bush
Buddleja davidii
FIGWORT FAMILY
Fl: 6-10 Ht: 1-5 m

This shrub of urban gap sites was introduced from China last century and has escaped from gardens at scattered sites around Scotland, including a few islands. It has *toothed, narrowly egg-shaped leaves* that are *white-felted beneath,* and *dense, pyramidal heads* of pollen-rich *lilac or purple flowers.* These have an *orange ring at the mouth* of a cylindrical petal-tube.

Hedge Woundwort
Stachys sylvatica
THYME FAMILY

Fl: 7-8 Ht: 30-100 cm

Hedge Woundwort is most easily recognised by its *pungent odour when crushed*. It has *roughly hairy, coarsely-toothed, heart-shaped leaves* and a spike of *claret-red, 2-lipped flowers, about 15 mm long*, with pale blotches on the lower lip. It is common in woods and hedgebanks over most of Scotland, but rare in the W Isles and absent from Shetland.

White Dead–nettle
Lamium album
THYME FAMILY

Fl: 5-12 Ht: 20-60 cm

This creeping perennial with *downy, toothed, heart-shaped leaves* spreads by underground runners to form *dense patches* in places. Its *creamy-white flowers* are *20–25 mm long*, with a *hooded upper lip* and a *ring of hairs* near the base of the petal-tube. It grows in hedgebanks, roadsides and wasteground in the lowlands and E of Scotland only.

Red Dead-nettle

Lamium purpureum

THYME FAMILY

Fl: 3-10 Ht: 10-45 cm

Found in cultivated and waste ground throughout Scotland, although more rarely in the Highlands and NW, this annual has *square stems*. Its leaves are *pungent, hairy and nettle-like but non-stinging*, with *rounded marginal teeth*. Its 2-lipped flowers are *pinkish-purple, 10–15 mm long*, with a long petal-tube and a *ring of hairs* near the base of the petal-tube.

Henbit Dead-nettle

Lamium amplexicaule

THYME FAMILY

Fl: 4-8 Ht: 5-25 cm

Henbit Dead-nettle grows in low-intensity cultivated land in the E lowlands and W coasts. It has *rounded, wavy-edged (rather than toothed) leaves*, the upper *stalkless ones seeming to encircle the stem*. Its *pinkish-purple flowers*, in whorls up the stem, have a *narrow petal-tube, much longer than the sepals*, with a hooded upper lip and *strongly 2-lobed lower lip*.

Common Hemp-nettle

Galeopsis tetrahit

THYME FAMILY

Fl: 7-9 Ht: 10-100 cm

Although similar to Dead-nettles, Hemp-nettles have 2 small bumps near the base of their petal-tubes. This annual is widespread in arable land and woods in lowland areas. It has *roughly-hairy stems* and *hairy, oval, toothed leaves*. Its *white, pink or purple flowers* often have dark blotches on the *3-lobed lower lip*, and the *petal-tube is no longer than the sepals*.

Lesser Burdock

Arctium minus

DAISY FAMILY

Fl: 7-10 Ht: 60-130 cm

As its name suggests, this plant has downy, *dock-like leaves* and *bur-like fruits*, about 2 cm in diameter, covered in *hooked bristles* which spread the fruits in fur and clothing. It is a sturdy biennial with arching stems and *egg-shaped, purple thistle-like flowerheads*. It grows in open woodland, roadsides and wasteground, although rarer inland in N Scotland and the N Isles.

DAISY FAMILY

Members of this family have tiny flowers, called **florets**, in a head resembling a single large flower, surrounded by sepal-like bracts. In many species, like Michaelmas Daisy (right), the petal-tubes of the outer florets are expanded into petal-like '**rays**' to attract pollinating insects. The inner '**disc**' florets are reproductive. Some species have only disc or ray florets.

Common Knapweed
Centaurea nigra
DAISY FAMILY
Fl: 6-10 Ht: 15-60 cm

This thistle-like plant of grassland and waysides is recognised by its tight heads of purple florets. These are borne on an oval cup, *2–4 cm in diameter*, of *dark brown bracts* which are cut into long, fine teeth at their tips. Its stems are rigid and its leaves are *gently-lobed*. It grows throughout Scotland, although less commonly in the N and introduced to Shetland.

THISTLES

Thistles (like Musk Thistle of E coastal grasslands, right) are recognisable by their spiny, lobed leaves and bulbous flowerheads, which are usually purple (although occasionally white in some species). These lack ray-florets, but consist of a cluster of deeply-divided tubular florets. The swollen flower base is clothed in sepal-like bracts which are often spiny.

Welted Thistle

Carduus crispus

DAISY FAMILY

Fl: 6-8 Ht: 30-120 cm

Confined mainly to wasteground and rough grassland in the E lowlands, Welted Thistle has gangling, branched, cottony *stems* with *spiny wings stopping just beneath the flowerheads.* The clustered, almost spherical heads, *broader and redder* than those of Marsh Thistle (p. 52), are surrounded by *spreading, narrow, green bracts* ending in a *weak spine.*

Creeping Thistle
Cirsium arvense
DAISY FAMILY
Fl: 7-10 Ht: 30-90 cm

Far-creeping roots make this a troublesome weed of fields and waysides around Scotland. Its *grooved* stems have *no leafy wings*, and it has *stalkless*, prickly leaves. Its inflorescences have *many, dullish-purple flowerheads*, up to 2.5 cm long. The flower base has *many, egg-shaped bracts pressed tightly around it*, but with *spreading, spine-tipped bracts at its foot*.

Spear Thistle
Cirsium vulgare
DAISY FAMILY
Fl: 7-10 Ht: 30-150 cm

Another abundant, widespread species, this is distinguished by its *spiny-winged, grooved, cottony stems*, *prickly-hairy leaves* with *spear-shaped, spiny lobes*, and *larger flowerheads* (to 5 cm long), which are *solitary or 2-3 in a group*. The *florets* are usually a *redder purple* and the *bracts* around the flower base are *narrow, spreading and spine-tipped*.

Marsh Thistle
Cirsium palustre
DAISY FAMILY

Fl: 7-9 Ht: 30-150 cm

The third common thistle (cf p. 51), this grows in marshes and damp grassland all around Scotland. Its cottony stem has a *continuous spiny wing*. Its leaves are *somewhat hairy above, narrowly-lobed, spine-tipped* and often *purple-bordered*. Its flowerheads, in *leafy clusters*, are *reddish-purple (or often white) and up to 2 cm long*, with purplish, pointed but *not spine-tipped bracts*.

Prickly Sowthistle
Sonchus asper
DAISY FAMILY

Fl: 6-8 Ht: 20-120 cm

This is marginally the commonest of 3 Sowthistles in Scotland. All are widespread plants of cultivated land and wasteground with rather spiny leaves, yellow dandelion-like heads and stems which exude a milky juice when cut. Prickly Sowthistle has *hairless stems* and *glossy-green leaves* with ·*spiny margins* and *rounded, back-projected basal lobes* (auricles).

Daisy

Bellis perennis

DAISY FAMILY

Fl: 2-10 Ht: 2-8 cm

Daisy is ubiquitous in short grassland and disturbed ground throughout Scotland, including all the islands. Its rosette of *glossy, spoon-shaped leaves* hugs the ground, escaping both grazing animals and the blades of lawn-mowers. Its abundant 'flowers', on hairy, leafless stalks have showy, *white or pink-tinged* outer ray-florets and a central disc of *yellow florets*.

Corn Marigold

Glebionis segetum

DAISY FAMILY

Fl: 6-8 Ht: 20-50 cm

A long-standing introduction from the Mediterranean, Corn Marigold has become rarer in arable fields due to modern herbicides. It is now commonest in croft fields in NW Scotland and the W Isles. It has yellow daisy flowers, *up to 6 cm across*, with spreading ray florets, and *toothed or lobed, bluish-green leaves*, at least the upper of which *half-clasp the stem*.

Scentless Mayweed
Tripleurospermum inodorum

DAISY FAMILY

Fl: 7-9 Ht: 50-60 cm

The commonest daisy-like weed of cultivated and waste ground, this annual is found across the lowlands, E coast and islands. Its typical daisy *flowerheads* are *4 cm across* with a *flat, yellow disc* and *white rays*. Its *leaves* are divided into *fine hair-like segments* (cf Ox-eye Daisy, p. 97). Sea Mayweed (*T. maritimum*) of coastal cliffs and dunes has *fleshier leaves*.

Pineapple-weed
Matricaria discoidea

DAISY FAMILY

Fl: 6-8 Ht: 5-30 cm

A successful introduction from Asia via North America, first recorded in Britain in 1871, Pineapple-weed grows in wasteground and trampled paths throughout Scotland. It has tufted leaves resembling Mayweed but with a *fruity smell when crushed*, and button-like, *greenish-yellow flowerheads*, *lacking ray florets* and less than 1 cm across.

Common Ragwort
Senecio jacobaea
DAISY FAMILY
Fl: 6-10 Ht: 30-150 cm

Common Ragwort is abundant in overgrazed pastures, wasteground, roadsides and sand-dunes throughout Scotland. It has *finely-divided rosette leaves*, which wither before flowering, similar stem leaves, and a *flat-topped* inflorescence of yellow daisy heads, *around 2 cm across*. These are enclosed by bracts, the *outer 2–5 of which are shorter than the rest*.

Oxford Ragwort
Senecio squalidus
DAISY FAMILY
Fl: 5-10 Ht: 20-40 cm

Originally from Sicily, this alien is now common on old walls and wasteground in the S, less common elsewhere but is still spreading, even to Shetland. It is distinguished from Common Ragwort by its bushier growth, shorter stems, and *more spreading inflorescence, usually with 13 ray florets*. The *outer bracts* clasping the base of the flowerhead are *black-tipped*.

Groundsel

Senecio vulgaris

DAISY FAMILY

Fl: 1-12 Ht: 8-45 cm

Groundsel flowers abundantly in wasteground, arable fields and gardens throughout Scotland, and is sparser only in the C Highlands. Its *yellow button heads* normally lack the spreading ray florets of other ragworts, and have *black-tipped bracts* as in Oxford Ragwort (p. 55). The *densely-clustered buds* and *mostly hairless leaves* are also characteristic.

Coltsfoot

Tussilago farfara

DAISY FAMILY

Fl: 2-4; Ht: 5-15 cm

In late winter, Coltsfoot produces *sturdy, scaly, purple stems,* topped by *solitary golden-yellow daisy heads.* Once its fruits are ripe, its *large, heart-shaped leaves* begin to develop, *edged with purple teeth* and *covered in a whitish down.* Coltsfoot is common in disturbed ground, tracks, dunes and screes throughout Scotland, but introduced to Shetland.

CARROT FAMILY

Members of this family (called **umbellifers**) typically have much-divided, lacy or fern-like leaves and tiny flowers clustered on stalks, which spread out like the spokes of an umbrella into convex or flat heads. They are best distinguished by leaf shape, stem form, the presence of **bracts** and **bracteoles** (see diagram) and the shape of the fruits, when present.

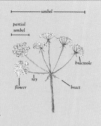

Rough Chervil
Chaerophyllum temulum

CARROT FAMILY

Fl: 6-7 Ht: 30-100 cm

Somewhat similar to Cow Parsley (p. 58), Rough Chervil is distinguished by its *rough, purple-spotted* stem, *swollen at the joints, softly hairy* leaves, and diffuse inflorescence, usually without bracts. Its *flask-shaped* fruit is 6 mm long, *narrowing at the tip* and often purplish. It grows in hedgerows and grassland in E Scotland, the Moray coast and a few sites in the SW.

Cow Parsley
Anthriscus sylvestris

CARROT FAMILY

Fl: 4-7 Ht: 60-100 cm

Cow Parsley becomes the dominant roadside umbellifer as Sweet Cicely (below) fades in June. It grows in hedgerows and bushy places over most of Scotland. It is distinguished by its *hollow, furrowed stem*, which is *hairy at the base*, and its *more finely-divided leaves*. Its inflorescence has *no bracts*, and its flowers ripen into *unridged, narrowly egg-shaped fruits*.

Sweet Cicely
Myrrhis odorata

CARROT FAMILY

Fl: 5-6 Ht: 60-100 cm

The first of three common roadside umbellifers to produce massed displays, before either Cow Parsley (above) or Hogweed (p.60), this downy perennial also grows in woods and grassland in the S and E. Its *fern-like* leaves are *white-blotched at their centres* and *smell of aniseed* when crushed. Its creamy flowers develop into *narrow, ridged, flask-shaped fruits*.

Ground-Elder
Aegopodium podagraria
CARROT FAMILY
Fl: 5-7 Ht: 40-100 cm

This troublesome perennial spreads by *far-reaching underground stems*. Its leaves have *3 lobes, often further divided into 3* and *toothed round the edge*. Its *stems are hollow and grooved*. Its inflorescence *lacks bracts and bracteoles*, and develops into *ridged, egg-shaped fruits*. It grows in cultivated land across Scotland, but not in parts of the C and N Highlands.

Wild Angelica
Angelica sylvestris
CARROT FAMILY
Fl: 7-9 Ht: 30-200 cm

Wild Angelica has a *robust, hollow, purplish stem*. Its leaves are divided into *narrow egg-shaped lobes*, and the *inflated base of the leaf-stalk wraps around the stem*. Its inflorescence has *hairy stalks* and *no bracts*, and its *flattened, egg-shaped fruits have broad, marginal wings*. It grows in damp meadows, roadsides, woods and marshes throughout Scotland.

Hogweed
Heracleum sphondylium
CARROT FAMILY
Fl: 7-9 Ht: 50-200 cm

The last of 3 common roadside umbellifers to flower (see p. 58), Hogweed grows in grassy places and hedgerows around Scotland, but more sparsely in the N. Its rough leaves are *once-divided* into broad, toothed lobes and *sheath the hollow stem* at their base. Its pinkish inflorescence is *flat-topped* and its *fruits* are *round, flattened and pale* (cf Giant Hogweed, p. 184).

Upright Hedge–Parsley
Torilis japonica
CARROT FAMILY
Fl: 7-8 Ht: 5-120 cm

Shorter and less bushy than other hedgerow and grassland umbellifers, this species is widespread in S and E Scotland, occasional in the W Isles and absent from far N and N Isles. Its *stem* is covered in *down-turned bristles* and feels *rough when rubbed upwards*. Its *flowers* are *pinkish or purplish-white*, and develop into *small, spiny, egg-shaped fruits*.

PLANTS OF GRASSLANDS AND MEADOWS

Many of Scotland's grassy hillsides are devoid of colour as a result of intensive agricultural methods and heavy grazing by sheep and deer. Only resilient and adaptable species like buttercups and thistles grow in many of these over-used pastures, while Daisy, Yarrow and Sheep's Sorrel are the only flowers to brighten many heavily-grazed hillsides in the Southern Uplands.

One remarkable plant that is spreading over these hillsides is Bracken (*Pteridium aquilinum*), a fern whose vegetation is poisonous to farm animals, causing gut cancer. Most animals know this and avoid eating it, allowing it to flourish while other plants are squeezed out by heavy grazing elsewhere. Trampling by cattle will knock back its tender young shoots, but cattle have largely been replaced on the Scottish hills by sheep, which are too light to keep this tough fern at bay. It continues to spread, apparently at the expense of grassland flowers.

Common Rock-rose grows only in sunny grassland on lime-rich rocks, as here in Holyrood Park, Edinburgh.

GRASSLANDS AND MEADOWS

Over most of Scotland, the underlying rocks are hard and impermeable granites, basalts, gneiss and other igneous rocks. These do not readily break down to release minerals into the soil, so produce soils that are acidic or mineral-poor. However, there are limited areas in Scotland where softer, more base-rich rocks enrich their overlying soils. Sometimes, these outcrops are highly localised, but they are worth seeking out for plants like Common Rockrose, Maiden Pink, Cowslip and Bloody Cranesbill.

Some grassland plants have adaptations to overcome the poor soil. Many members of the Pea Family, especially Clovers, have a fungus that lives in compartments called nodules in their roots. The fungus is able to take nitrogen from the atmosphere and make nitrates. It supplies some of these to its host plant, perhaps as reward for the protection it gets.

Several other grassland plants live as 'partial parasites'. Their roots tap into the roots of other plants (often grasses) and extract some of the minerals they need for healthy growth. They do not 'steal' the sugars they require for energy, because they can make these themselves through photosynthesis (so they are not full parasites). This strategy allows species such as Yellow Rattle, Red Bartsia and Eyebrights to flourish in soils that might otherwise be too poor to support their growth.

Eyebrights are an interesting group. They are divided into a large number of very similar, closely related 'microspecies'. Some are species of herb-rich hay meadows, while others grow on coastal rocks and clifftop turf in the far north, including several that are found nowhere else in the world. Appropriately named, one of these, *Euphrasia campbelliae*, is found only in coastal turf on the Isle of Lewis.

One Scottish habitat that has disappeared almost completely is wet meadowland. Once, such

Melancholy Thistle is a northern species which flowers in July in a few remaining damp meadows. Photo: Sue Scott

*Clustered Bellflower (*Campanula glomerata*) is an uncommon species of lime-rich E coast sites.*

meadows were common in marshes and along riverbanks, but drainage and farm improvements have converted most of them to drier, more agriculturally productive grassy swards.

Wet meadows were the home of plants such as Globeflower, Wood Cranesbill and Melancholy Thistle. Today, these survive only in the few remaining fragments of meadowland, and in the 'tall herb' community of mountain rock ledges, beyond the reach of deer and sheep. Fortunately, a precious few 'left-over' meadows remain to remind us of former glories, several of which are now protected as nature reserves.

Globeflower
Trollius europaeus

BUTTERCUP FAMILY

Fl: 6-8 Ht: 10-60 cm

Buttercup-like from a distance, Globeflower is distinguished by its *globe-shaped flowers*, to 3 cm across, which are *completely enclosed* by 5–15 golden-yellow petal-like sepals. Its roundish leaves are *deeply cut into 3–5 spreading, toothed and divided lobes*. It forms occasional patches in hill meadows in the Highlands and Borders, especially by rivers.

Meadow Saxifrage
Saxifraga granulata

SAXIFRAGE FAMILY

Fl: 6-8 Ht: 10-50 cm

The *round-toothed, kidney-shaped, hairy leaves* of this saxifrage are up to 3 cm across and mostly in a basal rosette. They have *pinkish bulbils* (small buds) at their base. The few-leaved, *softly-hairy stem* is topped by a spreading head of white flowers, *up to 2 cm across*, with *broad, rounded petals*. It grows uncommonly in open grassland in S and E Scotland.

THE PEA FLOWER

Flowers of the pea family, like those of Common Birdsfoot-trefoil (right), characteristically have a broad upper petal (the '**standard**'), 2 narrower side petals ('**wings**'), and 2 lower petals united at their base to form a boat-like '**keel**'. Pollination is by heavy insects, such as bees, which land on and weigh down the keel to reach the hidden pollen.

Common Birdsfoot-trefoil

Lotus corniculatus

PEA FAMILY

Fl: 6-9 Ht: 10-40 cm

A *hairless* perennial of grasslands throughout Scotland, Birdsfoot-trefoil actually has *5 leaflets*, but the lowest 2 are *attached to the stem*, like stipules, leaving 3 free (trefoil) lobes. Its *solid stems* creep along the ground, forming spreading patches, with heads of 2–6 yellow pea flowers on stout flower-stalks, which develop into pods which *spread like birds' feet*.

Tufted Vetch
Vicia cracca
PEA FAMILY

Fl: 6-8 Ht: scrambling to 2 m

Vetches are scrambling plants, with leaves divided into many pairs of opposite leaflets and ending in a twining tendril. This species has a *one-sided, densely-flowered spike of up to 40 blue-violet flowers, 8–12 mm long*, and *slightly hairy leaves with 6–12 pairs of narrow oval leaflets* and a *branched tendril*. It grows in grassy or bushy places round most of Scotland.

Bush Vetch
Vicia sepium
PEA FAMILY

Fl: 5-8 Ht: scrambling to 1 m

Reminiscent of the previous species, this is distinguished by its *hairless leaves* which have *5–9 pairs of broad egg-shaped leaflets* ending in a *branched tendril*. Its inflorescence is shorter (*1-2 cm long*) and less showy, with *2–6, pale purple flowers* which are *12–15 mm* long. It is widespread in rough grassland, hedges and thickets around virtually all of Scotland.

Common Vetch
Vicia sativa
PEA FAMILY
Fl: 5-9 Ht: scrambling to 1.2 m

This very variable, tufted, slightly hairy, sprawling *annual* is found in hedges, grassy places and sand-dunes around the lowlands and NE, but not in most of the Highlands or Islands. Its leaves have *3–8 pairs of narrow leaflets* with a *branched or unbranched tendril* and its *reddish purple flowers*, up to *2 cm long*, are usually *solitary or paired* at the base of a leaf.

Bitter Vetch
Lathyrus linifolius
PEA FAMILY
Fl: 4-7 Ht: 10-40cm

Peas and Vetchlings (*Lathyrus* species) differ from Vetches in having *winged or angled stems* and *fewer pairs of leaflets*. Bitter Vetch(ling) has *reddish-purple flowers, hairless leaves* with *2–4 pairs of lance-shaped leaflets* and no tendrils. At the base of its leaf stalks it has *lance-shaped* flaps (stipules). It grows in woods, hedgebanks and heaths throughout Scotland.

Meadow Vetchling
Lathyrus pratensis
PEA FAMILY

Fl: 5-8 Ht: 30-120 cm

This is the commonest vetchling (cf p. 69) in grassland and hedges throughout Scotland, although rare in the NW. It has *yellow flowers*, in clusters of *5–12*, and scrambling, *sharply-angled* but *unwinged stems*. Its *downy leaves* have a *single pair* of *lance-shaped leaflets*, a climbing *tendril* and a pair of *leafy, arrow-shaped flaps* (stipules) at the base of the leaf stalk.

Common Restharrow
Ononis repens
PEA FAMILY

Fl: 6-9 Ht: 30-60 cm

'Restharrow' records how the underground stems once brought horse-drawn harrows to a halt. This is the only widespread Scottish species, found in rough grassland and sand-dunes in the E lowlands and near W coasts. It has *hairy, shrubby stems*, sometimes with soft spines, leaves with 3 rounded, *downy, toothed leaflets*, and *pink flowers*, 10–15 mm long.

Black Medick
Medicago lupulina
PEA FAMILY
Fl: 5-9 Ht: 5-50 cm

This low, downy annual or perennial has a similar range to Restharrow (opposite) in grassy places and roadsides. Its leaves have *3 broad leaflets* ending in a *minute point*. In summer, it has round, compact heads of 10–50 *tiny, deep yellow flowers* (about 3 mm long). These develop into *kidney-shaped pods*, up to 3 mm long, which turn black when ripe.

Lesser Trefoil
Trifolium dubium
PEA FAMILY
Fl: 5-10 Ht: trailing to 25 cm

The commonest *yellow-flowered* clover (see p. 72) all round Scotland, this annual has *long-stalked, rounded heads*, about *7 mm across*, of 3–15 tiny flowers, in the angles of its upper leaves. Its leaves have 3, *oval leaflets* and *oval stipules*. It resembles Black Medick (above), but has *hairless* (not downy) *leaves* and sepals, and its *leaflets* are *slightly notched at the tip*.

White Clover
Trifolium repens
PEA FAMILY
Fl: 6-9 Ht: up to 50 cm

Clovers have *trefoil leaves* (i.e. with 3 leaflets) and *dense heads* of small, stalkless flowers with *wings longer than their keel* (see pea flower, p. 67). White Clover is abundant in grassy places throughout Scotland. It has *creeping, rooting stems*, toothed *leaflets*, usually with a *pale band* towards their base, and round flowerheads with 40–80 *white or pale pink flowers*.

Red Clover
Trifolium pratense
PEA FAMILY
Fl: 5-9 Ht: up to 60 cm

Widespread in Scottish grasslands, Red Clover *lacks creeping stems* and has *globe-shaped* heads of *pinkish-purple flowers*. Its leaves have *narrowly oval leaflets*. They are usually marked with a *whitish crescent* and are *hairy beneath*, with *oblong stipules*. Zigzag Clover (*T. medium*), with *narrower leaflets* and *flat heads* of *reddish-purple flowers*, is less common.

Haresfoot Clover
Trifolium arvense

PEA FAMILY

Fl: 6-9 Ht: 5-20 cm

The flowers of this softly hairy annual have sepals ending in *long, bristly hairs*, which give the inflorescence a *fluffy appearance*. The sepal bristles completely hide the *white or pink flowers*, which are only 4 mm long. Its leaves have *narrow, hairy leaflets*. It inhabits sandy fields, grassland and dunes, mostly near coasts, in S Scotland and the Moray Firth area.

Silverweed
Potentilla anserina

ROSE FAMILY

Fl: 5-8 Ht: creeping to 80cm

The only common yellow flower with *silvery compound leaves*, Silverweed grows in damp grassland, sand-dunes and roadsides in all but the C Highlands. It spreads by *creeping, rooting stems*, producing regular rosettes of leaves, to 25 cm long, with *7–12 pairs of toothed leaflets*, which are *silky-hairy beneath*. Its 5-petalled flower is 2 cm or more across.

Tormentil
Potentilla erecta
ROSE FAMILY
Fl: 6-9 Ht: 5-50 cm

A perennial of grassland, heaths, bogs, woods and mountainsides throughout Scotland, Tormentil has yellow, *4-petalled* (occasionally 3- or 5-petalled) *flowers, less than 1 cm across*, and leaves with 3 (occasionally 4 or 5) *spreading, toothed leaflets*. Its slender, flexuous stems arise from a thick, creeping, woody stock, but *do not root along their length*.

Common Lady's-mantle
Alchemilla vulgaris
ROSE FAMILY
Fl: 6-9 Ht: 5-45 cm

The name is used here to cover a group of 10 rather similar perennials found in grassland and open woods throughout Scotland except the far NW. All have leaves that are round in outline, pleated like a cloak when young, and with shallowly-cut, toothed lobes. The tiny flowers, in *dense clusters* in summer, have *no petals* but 4, greenish-yellow sepals.

74

Fairy Flax
Linum catharcticum
FLAX FAMILY
Fl: 6-9 Ht: 5-25 cm

This slender annual is found virtually throughout Scotland, although nowhere abundantly, in grassland, heaths and dunes, usually on base-rich soils. It has *unbranched, wiry stems* and *opposite, oblong leaves*. Its small *flowers, about 8 mm across*, are borne in *widely-branched inflorescences*. They have 5 sepals, *5 narrow, white petals* and 5 stamens.

Perforate St John's-wort
Hypericum perforatum
ST JOHN'S-WORT FAMILY
Fl: 6-9 Ht: 30-90 cm

This is the commonest of several similar perennials, all with opposite, oval leaves and clusters of yellow, 5-petalled flowers with *numerous stamens*. This species is distinguished by its *narrow leaves peppered with translucent, glandular dots* (best seen against the light) and stems with *2 raised lines*. It grows in grassland, woods and hedgebanks in the S and E.

VIOLET FAMILY

Violets and pansies (like Heath Dog-violet; right) have toothed, usually heart-shaped leaves and flowers borne singly on long stalks. The flowers have 2 top petals projecting upwards, 2 equally-sized side petals and an enlarged lower petal, which is often marked with lines guiding insects to nectar in the long tube-like **spur** which extends behind the lowest petal.

Common Dog-violet

Viola riviniana

VIOLET FAMILY

Fl: 4-6 Ht: 2-20 cm

Common in woods, hedgebanks, heaths, and grassland throughout Scotland, this perennial is distinguished by its *generally hairless, heart-shaped leaves, and blue-violet* flowers with a *paler spur,* notched at its tip. Heath Dog-violet (*V. canina*) (see above) with *bluer flowers,* a *greenish-yellow spur* and *narrow triangular leaves* grows in scattered heaths and fens.

Wild Pansy

Viola tricolor

VIOLET FAMILY

Fl: 4-9 Ht: up to 15 cm

This annual or perennial grows in grassland and cultivated ground around most of Scotland, except for the C and W Highlands. It resembles Field Pansy (p. 22) but has larger flowers (*to 25 mm vertically*), with *petals longer than the sepals* and coloured any combination of *purple, yellow and white*. The *spur is twice the length of the flaps at the base of the sepals.*

Bloody Cranesbill

Geranium sanguineum

CRANESBILL FAMILY

Fl: 5-8 Ht: 10-40 cm

This striking Cranesbill has *rich purplish-crimson* flowers, which are *up to 3 cm across*, with *shallowly notched petals*, and are usually *solitary* on long stalks. It has a rather *bushy form* with spreading, hairy stems and leaves *deeply cut into many narrow* lobes. It grows uncommonly on lime-rich rocks and sand-dunes, mostly near the coast in S and E Scotland.

Common Rock-rose
Helianthemum nummularium
ROCK-ROSE FAMILY
Fl: 6-9 Ht: 5-30 cm

The only Scottish member of a mostly Mediterranean family, Common Rock-rose grows on grassy lime-rich banks in S and E Scotland (see p. 62). A low-growing *shrub* with trailing stems, it has *opposite pairs of narrow, oval leaves* with a *white down* underneath. Its showy, *shining yellow*, 5-petalled flowers are about 2 cm across, *with many stamens*.

Common Whitlow-grass
Erophila verna
CABBAGE FAMILY
Fl: 3-6 Ht: 2-20 cm

This tiny, short-lived, spring-flowered annual grows in dry, open grassland, rocks and dunes, but is scarce in the N and absent from the N Isles. It has a *basal rosette of lance-shaped* leaves, a short, leafless flowering stem and a branched inflorescence of white *flowers, up to 6 mm across*, with 4 *deeply-notched petals*. These ripen into *flattened, oval fruit-pods*.

Sheep's Sorrel
Rumex acetosella

DOCK FAMILY

Fl: 5-9 Ht: up to 30 cm

This slender, creeping docken (p. 34) grows in heathy and grassy places throughout Scotland, often on poor soils. Its narrow, *arrow-shaped leaves*, which are up to 4 cm long, have *upward-pointing lobes* at their base, and all are *distinctly stalked*. The reddish flowers are borne in well-spaced clusters in a leafless, slightly branched inflorescence.

Common Sorrel
Rumex acetosa

DOCK FAMILY

Fl: 5-6 Ht: up to 50 cm

Even more common in grassland, hedgebanks and woodland clearings than Sheep's Sorrel, this is distinguished by its *broader, lance-shaped leaves, to 15 cm long*, with *downwardly-directed basal lobes*, and upper leaves *clasping the stem*. The *flower-spikes* are *broader* than those of Sheep's Sorrel, and most are in fruit by August, when the plant turns rich crimson.

Maiden Pink
Dianthus deltoides

PINK FAMILY

Fl: 6-9 Ht: 15-45 cm

This delicate perennial grows on lime-rich grassy banks only in the E, but is declining through over-grazing and scrub encroachment. Its *narrow, blue-green, rough-edged leaves* are easily overlooked amongst tall grass, despite its attractive flowers. These are *blushing pink*, usually with a *pale centre surrounded by a darker pink band*, and with *petals frayed at their tips*.

Cowslip
Primula veris

PRIMROSE FAMILY

Fl: 4-6 Ht: 10-30cm

Cowslip is a decreasing perennial of *base-rich* pastures, mostly in E Scotland, with a few sites in the W, N coast and Orkney. It is related to Primrose (p. 138), but has up to 30 *nodding, deep-yellow flowers* in a spreading head on a *leafless flower-stalk*. Its *tubular flowers* have *short, inward-curved petal-lobes*. Its *gently-toothed* leaves *narrow abruptly* into a

leaf-stalk,
Lady's Bedstraw
Galium verum

BEDSTRAW FAMILY

Fl: 7-8 Ht: 15-80 cm

The commoner of 2 yellow-flowered Bedstraws (see below), this species grows in grassland, hedgebanks, machair and sand-dunes throughout Scotland, although less commonly in the NW. It has *sprawling, 4-angled stems* and dense, fluffy heads of *golden-yellow flowers*. Its *narrow, needle-like leaves*, in whorls of 8–12, are usually *hairless* with incurled margins.

Crosswort
Cruciata laevipes

BEDSTRAW FAMILY

Fl: 5-6 Ht: 15-70 cm

The only other yellow-flowered Bedstraw, Crosswort has *broader, softly hairy* leaves, with *3 prominent veins*, arranged in *cross-like whorls of 4* up its scrambling stems. Its 4-petalled flowers, *2–3 mm in diameter*, are *tightly clustered into the angles of its upper leaves (bracts)*. It is a plant of lime-rich grassland, roadsides and open woodland, mainly in S Scotland.

Field Gentian
Gentianella campestris
GENTIAN FAMILY
Fl: 7-10 Ht: 10-30 cm

Field Gentian grows in pastures and dunes, mostly in the N and W. It is a hairless biennial with a slightly-branched stem and lance-shaped leaves. Its flowers have a *bluish-lilac (or occasionally white) petal-tube*, about 2 cm long, with 4 out-spread lobes. The *petals just overtop the sepal-tube*, which has *2 broad outer lobes almost hiding 2 narrower inner lobes.*

Autumn Gentian
Gentianella amarella
GENTIAN FAMILY
Fl: 8-10 Ht: 5-30 cm

Autumn Gentian has *deeper purple* flowers than Field Gentian (or occasionally pink, white or bluish). Its petal-tube usually has *5 lobes* (sometimes with 4-lobed flowers on the same plant) about *twice as long as the sepal-tube* which has *4 or 5 equally-sized teeth*. It grows in lime-rich grassland around coasts and islands, and inland in the C and NW Highlands.

Viper's Bugloss
Echium vulgare
BORAGE FAMILY
Fl: 6-9 Ht: 30-90 cm

This handsome biennial grows on sea-cliffs, dunes, and light soils in the S, with scattered introductions in the N. Its roughly *hairy*, lance-shaped leaves have a *prominent midrib*, and the upper ones *clasp the stem*. It has spikes of funnel-shaped *blue flowers* (pink in bud) with a *prominent lower lip*, and 1 or more *pink stamens projecting from the flower's throat*.

Changing Forget-me-not
Myosotis discolor
BORAGE FAMILY
Fl: 5-9 Ht: 8-25 cm

Whereas most Forget-me-nots have pink buds, the flowers of this annual are yellow when they open and turn blue, so *young flowers near the tip of the curled inflorescence are yellow* with blue flowers lower down. It has *hairy stems and leaves*, and *flower-stalks shorter than its sepal-tubes*. It grows in open grassland throughout Scotland, but is not very common.

SPEEDWELLS

17 speedwells are found in Scotland. 2 species live in the mountains, 4 in wet areas and 11 in lowland grassland, like Ivy-leaved Speedwell (right). All have petal-tubes with 4 lobes, of which the uppermost is largest (formed from 2 united petals); the lowest is often narrower. Two stamens and 1 style protrude from the petal-tube mouth. Most have blue flowers.

Heath Speedwell
Veronica officinalis

SPEEDWELL FAMILY

Fl: 5-8 Ht: 10-40 cm

This creeping, *mat-forming*, downy perennial is widespread in heaths, grassland and open woods throughout Scotland, including the Isles. Its leaves are *oval and hairy*, and its flowers are *small* and *pale lilac*, in a *dense, leafless spike* emerging from the angle of stem leaves. As in many Speedwells, prominent blue lines on the flowers guide insects to their pollen.

Thyme-leaved Speedwell
Veronica serpyllifolia

SPEEDWELL FAMILY

Fl: 3-10 Ht: 10-30 cm

Common throughout Scotland, this perennial grows in damp grassland, disturbed ground and heaths. Its *flowers are white or pale blue with darker lines*, borne in the angle of the upper stem leaves, which are *oblong and hairless*. Ivy-leaved Speedwell (*V. hederifolia*) (box opposite), with similar flowers but *kidney-shaped leaves*, grows in cultivated ground in the E.

Common Field Speedwell
Veronica persica

SPEEDWELL FAMILY

Fl: 2-11 Ht: 10-40 cm

This *annual* produces *bright-blue flowers* with a *paler bottom lobe* in lowland fields and wasteground around Scotland. It has *hairy, rather straggling stems* and *egg-shaped leaves*. The rather similar Wall Speedwell (*V. arvensis*) is distinguished by its *tiny, bright-blue flowers* (*rarely exceeding 3 mm across*). It grows in open or cultivated ground.

Germander Speedwell

Veronica chamaedrys

SPEEDWELL FAMILY

Fl: 3-7 Ht: 20-40 cm

This Speedwell forms sprawling patches in grassland and woods throughout Scotland. It has *toothed, egg-shaped leaves*, and *2 lines of white hairs down opposite sides* of its weak stem. Its flowers are *deep blue* with a *prominent white 'eye'*, on long stalks. Slender Speedwell (*V. filiformis*), a garden escape, forms sheets of *pale lilac-blue* flowers in lowland grassland.

Ribwort Plantain

Plantago lanceolata

PLANTAIN FAMILY

Fl: 4-8 Ht: up to 45 cm

Plantains have a basal rosette of ribbed or veined leaves, leafless flowering stems, and dense spikes of flowers with reduced petals and sepals but prominent stamens. Ribwort Plantain, an abundant species of grassy places throughout Scotland, has *lance-shaped leaves*, 2–30 cm long, with *3–5 prominent veins*, and a *dumpy, egg-shaped inflorescence*.

Greater Plantain
Plantago major
PLANTAIN FAMILY
Fl: 5-8 Ht: 10-25 cm

Greater Plantain grows in disturbed ground, including lawns, gardens, fields, roadsides and wasteground, throughout Scotland. It has *broader, oval, usually hairless leaves*, with *3–9 veins*. The leaves are 5–15 cm long, with *stalks as long as their blades*. The narrow *inflorescence* is 10–15 cm long and takes up most of the *unfurrowed stem* (cf pp 86, 121).

Bugle
Ajuga reptans
THYME FAMILY
Fl: 5-7 Ht: 10-30 cm

Somewhat reminiscent of Selfheal (p. 88), Bugle has *hairless, scarcely toothed* leaves. The *upper leaves and bracts* are *almost stalkless* and often purple, and its unbranched *stems* are *usually hairy on 2 sides*. It has dense, leafy spikes of *violet-blue* flowers with a *strongly 3-lobed lower lip*. It grows in damp grassland and woods, less commonly in the N, and is absent from Shetland.

Ground Ivy
Glechoma hederacea
THYME FAMILY

Fl: 3-6 Ht: 10-30 cm

This *hairy*, patch-forming perennial is common in grassland, woodland and hedgebanks in S and E Scotland, but absent from much of the NW and W Isles, and a garden outcast in the N Isles. It has *creeping, rooting stems* with *kidney-shaped, roundly-toothed* leaves. Its *blue-violet* flowers, in whorls of 2-4, have lower lips which are *spotted with deeper purple*.

Selfheal
Prunella vulgaris
THYME FAMILY

Fl: 6-9 Ht: 5-30 cm

The *oval, slightly toothed leaves* of this patch-forming, *downy* perennial are found in lawns, grassland and woodland clearings throughout Scotland. Its *squarish* flowering stems are topped by *dense, oblong heads* of *violet, 2-lipped* flowers (rarely pink or white), surrounded by *leafy, purplish bracts*. The lower flower-lip is 3-lobed and the upper strongly hooded.

Wild Thyme
Thymus polytrichus
THYME FAMILY

Fl: 5-8 Ht: up to 7 cm

This *mat-forming shrub* grows in dry grassland and heaths throughout the country. It has squarish *stems, with hairs on 2 sides,* and *opposite, leathery, oval leaves* which are *aromatic* when crushed. Its *rosy-purple* flowers grow in *domed heads.* The lower lip of their petal-tube is *strongly 3-lobed,* and the upper *slightly 2-lobed* with 4 protruding stamens.

Eyebright
Euphrasia officinalis
BROOMRAPE FAMILY

Fl: 5-9 Ht: 2-40 cm

Eyebrights are a group of many, near-identical 'microspecies'. All are low-growing grassland annuals, with *toothed, oval leaves,* and spikes of *white or purplish flowers* with a *4-toothed sepal-tube* and a *2-lipped petal-tube.* The upper petal lip is *curved backwards* and often purple-lined, and the lower lip is *strongly 3-lobed* and *blotched with yellow* near the throat.

Red Bartsia
Odontites vernus
BROOMRAPE FAMILY
Fl: 6-8 Ht: up to 50 cm

Red Bartsia grows in grassland in S Scotland and near coasts in the N and Isles. It is a *downy, often purple-tinted* annual, with *toothed, spear-shaped leaves*. Its *purplish-pink* flowers are about 8 mm long, all *bent towards one side* of a *leafy spike*. Its sepal-tube is *4-toothed* and its *petal-tube* has two lips, the upper of which forms an *open hood* and the lower is *3-lobed*.

Yellow Rattle
Rhinanthus minor
BROOMRAPE FAMILY
Fl: 5-9 Ht: 5-50 cm

This variable annual grows in grassland all round Scotland. It has often *black-spotted* stems, opposite pairs of *narrow, toothed, strongly-veined leaves*, and *leafy spikes* of yellow flowers. These are 15–20 mm long, with an *inflated sepal-tube* with 4 teeth, and a 2-lipped petal-tube. The upper lip is *flattened sideways, enclosing 4 stamens*, and the lower lip is *3-lobed*.

Scottish Bluebell
Campanula rotundifolia
BELLFLOWER FAMILY
Fl: 7-9 Ht: 15-40 cm

Although 'officially' called Harebell, this hairless perennial is usually known as Bluebell in Scotland. It grows in dry grassland, sand-dunes, machair and rock-ledges, but is less common in the NW and rare in the N Isles. It has *round, long-stalked root leaves* and *narrower, grass-like* stem leaves. Its *blue* flowers are *nodding and broadly bell-shaped*.

Sheep's-bit
Jasione montana
BELLFLOWER FAMILY
Fl: 5-8 Ht: 5-50cm

This *downy* biennial, with *narrow, untoothed leaves* grows in grassland, heaths and sea cliffs, mostly in the SW, rarer in the E, but commonly on Shetland. Its *small, blue* bell-flowers are deeply divided into *5 narrow lobes*, giving a tufted appearance to its *crowded flowerheads*. It is separated from scabiouses (p.98) by anthers *joined at the base into a short tube*.

Melancholy Thistle
Cirsium heterophyllum

DAISY FAMILY

Fl: 7-8 Ht: 45-120 cm

The only Scottish thistle *without spiny leaves*, this handsome perennial (see p. 64) grows in pastures and streamsides in hill districts. It has *grooved, cottony stems* and *soft, bright green leaves* with a *dense white felt underneath*. These may be *oblong and toothed* or deeply cut into *broad lobes*. Its purple flowerheads are 3–5 cm across on an egg-shaped cup of spiny bracts.

Nipplewort
Lapsana communis

DAISY FAMILY

Fl: 7-9 Ht: 20-90 cm

This slender annual grows in waysides and hedgerows around Scotland, but less commonly in the NW and Isles. It is distinguished from other dandelion-like plants by its *fruits*, which have *no parachute of hairs*. Its lower stem leaves are *lobed*, with a *large end lobe*, and its upper leaves are less divided. Its flowerheads are 2 cm across in a *branching inflorescence*.

Dandelion
Taraxacum officinale
DAISY FAMILY

Fl: 3-10 Ht: 5–30 cm

Dandelions are a group of over 230 closely-related, in-breeding 'microspecies', a good number of which are endemic to Scotland. They have *solitary flowerheads* on *unbranched, leafless stems* which *exude milky juice* when cut. The *leaves*, in a *basal rosette*, are generally oblong and *deeply lobed*, with the lobes *curving downwards* like lions' teeth. The yellow flowerheads have only ray florets (see daisy flower on p. 49) and are 1.5–5.5 cm across, often with *down-curved bracts*. The *beaked fruits* have a *tuft of hairs* for dispersal. Dandelions grow in grassland, wasteground and marshes throughout Scotland.

While Dandelions are familiar enough, there are many Dandelion-like plants which seem superficially similar, and are initially off-putting in their sheer variety. There is space for only the commonest of these species in this Mini-Guide. However, with practice and confidence, they can generally be separated by the shape of their leaves, inflorescence and bracts.

Cat's-ear
Hypochaeris radicata
DAISY FAMILY

Fl: 6-9 Ht: 20-60 cm

Widespread in grassland, waysides and dunes, but introduced on Shetland, Cat's-ear has a *basal rosette* of *hairy, toothed or lobed leaves*. Its *swollen, hairless stems* are usually *forked* and *covered in scale-like bracts* beneath the flowerheads. These are 2–4 cm across, with *pointed scales between the florets*. The outer *ray-florets are greenish or greyish below*.

Autumn Hawkbit
Scorzoneroides autumnalis
DAISY FAMILY

Fl: 6-10 Ht: 5-60 cm

Autumn Hawkbit grows in pastures and waysides across Scotland. It is distinguished from Dandelions by its *slender, branched stem*, and from Cat's-ear by *shiny hairless leaves*. Its stems are *little swollen*, with *scale-like bracts* below the flowerheads. These are 1.5–3.5 cm across, with a *woolly* cup of *bracts* at their base. The *outer ray-florets are reddish beneath*.

Smooth Hawksbeard
Crepis capillaris
DAISY FAMILY

Fl: 6-9 Ht: 20-90 cm

The *wiry stems* of this annual are *hairy at the base*. Its *few stems leaves* are *lance-shaped, lobed or toothed*. The upper leaves have a *clasping, arrow-shaped base*. The flowerheads, in an *open inflorescence*, are *1–2 cm across* and the outer florets are often *reddish underneath*. It grows in grassland, heaths and wasteground around Scotland, but not native on Shetland.

Mouse-eared Hawkweed
Pilosella officinarum
DAISY FAMILY

Fl: 5-8 Ht: 5-30 cm

Hawkweeds (*Hieracium* and *Pilosella* species) are a group of many 'microspecies', found in grassland all round Scotland. They are perennials with *hairy stems, hairy leaves* which are *unlobed or slightly toothed*, and yellow or orange-red dandelion-like flowers. This species has *only rosette leaves*, which are *white-felted beneath*, and *solitary flowerheads*.

Yarrow
Achillea millefolium
DAISY FAMILY
Fl: 6-9 Ht: 8-60 cm

One of the commonest grassland species throughout Scotland, Yarrow is recognised by its *feathery leaves* which are lance-shaped in outline but *deeply cut into 'thousands' of fine segments*. Its *flattish, umbrella-like inflorescence*, on top of a *woolly stem*, has flowerheads, about 5 mm across, with *usually 5 broad, white or pink rays* and a *creamy-white* disc.

Sneezewort
Achillea ptarmica
DAISY FAMILY
Fl: 7-8 Ht: 20-60 cm

The flowers show Sneezewort to be a relative of Yarrow, but its *flowerheads are bigger* (to 18 mm across), with *8–12 creamy-white rays* and a *greenish-white disc*. Its inflorescence is *more open and fewer-flowered*. The *leaves are lance-shaped, undivided but saw-edged*. Sneezewort grows, less commonly, in damp grassland and marshes throughout Scotland.

Ox-eye Daisy
Leucanthemum vulgare

DAISY FAMILY

Fl: 5-9 Ht: 20-70 cm

Called Gowans in parts of Scotland, Ox-eye Daisy is common in grassland in the S and more scattered in the N and Isles. It somewhat resembles Scentless Mayweed (p. 54), but its *leaves* are *toothed,* not finely divided. Its solitary daisy flowerheads, to 5 cm across, have white, petal-like rays, a yellow central disc, and oblong *bracts with narrow purple margins.*

Common Valerian
Valeriana officinalis

VALERIAN FAMILY

Fl: 6-8 Ht: 20-150 cm

This *strong-smelling* perennial grows in rough grassland, scrub and marshes over most of Scotland, but is sparser in the N and absent from Shetland. Its opposite leaves are divided into *several pairs of oval, toothed leaflets* with a single end leaflet. It has an umbrella-like cluster of *pale-pink* flowers with petals united into a *funnel-shaped* tube about 5 mm across.

Field Scabious
Knautia arvensis

TEASEL FAMILY

Fl: 7-9 Ht: 25-100 cm

A *roughly hairy* perennial of dry pastures, mostly in SE Scotland, Field Scabious has most of its stem leaves *deeply-cut* into narrow lobes. Its small flowers, in *dense, rounded heads*, are about 3 cm in diameter. The *bluish-lilac petals* are united into a 4-lobed tube, with the *outer lobes* of the enlarged outermost flowers forming a *frill round the inflorescence*.

Devil's-bit
Succisa pratensis

TEASEL FAMILY

Fl: 6-10 Ht: 15-100 cm

Also called Devil's-bit Scabious, this species is generally smaller than Field Scabious. It has *undivided, narrowly oval*, often purple-blotched *leaves*. In summer and early autumn, it produces rounded heads, about *2 cm in diameter*, of *dark purplish-blue flowers, without any frill* of outer petal-lobes. It grows in meadows, woods and marshland throughout Scotland.

Common Twayblade
Neottia ovata

ORCHID FAMILY

Fl: 6-8 Ht: 20-60 cm

Easily overlooked despite its sturdy stems, this orchid grows in base-rich grassland, open woods and dunes in the S and near coasts in the N and Isles (except Shetland). It has a *single pair of broadly oval, ribbed leaves, 5–20 cm long*. Its *yellowish-green flowers*, in a *loose spike*, have *hooded upper lobes*, 2 side lobes like arms, and a *long, deeply forked lower lip*.

Greater Butterfly Orchid
Platanthera chlorantha

ORCHID FAMILY

Fl: 5-7 Ht: 20-40 cm

This handsome orchid usually has a *single pair of oblong basal leaves*. Its *greenish-white flowers* have 3 spreading upper lobes, and a *strap-shaped lower lip*. The *down-curved spur* behind the flower is *19–28 mm long*. It is uncommon in woods and lime-rich grassland, mostly in C and W Scotland. Lesser Butterfly Orchid (*P. bifolia*), on poorer soils, has *an almost horizontal spur less than 20 mm long*.

Small White Orchid
Pseudorchis albida
ORCHID FAMILY
Fl: 6-8 Ht: 10-30 cm

This delicate orchid has a few *oblong, glossy lower leaves* and 1 or 2 bract-like stem leaves. It has a *cylindrical spike* of *greenish white, strongly hooded, bell-shaped flowers* with a *short, 3-lobed lower lip* and a *stumpy, down-curved spur*. It grows in rough hill pastures and open moorland in the Highlands and NW mainland, and, more rarely, in the Borders and Orkney.

Heath Fragrant Orchid
Gymnadenia borealis
ORCHID FAMILY
Fl: 6-8 Ht: 15-40 cm

Fragrant Orchids have recently been divided into 3 species. Their distribution is poorly known, but this is the commonest in Scotland, on hill pastures and moors. It has a *dense, cylindrical spike* of *rosy-pink to whitish, strongly fragrant flowers*, covered in *glistening scales*. The flowers are *strongly hooded*, with *slender, down-curved spurs*.

Common Spotted Orchid

Dactylorhiza fuchsii

ORCHID FAMILY

Fl: 6-8 Ht: 10-50 cm

One of the Marsh Orchid group, this grows in lime-rich grassland in the S and in machair and limestone rocks in the W. It has 5–12 *dark-blotched leaves,* and pale pink flowers marked with *red spots and dashes,* in a dense, cylindrical spike. The flowers resemble other marsh orchids, but have the lower lip *divided into 3 triangular lobes, the middle of which is longest.*

Early Purple Orchid

Orchis mascula

ORCHID FAMILY

Fl: 4-6 Ht: 15-60 cm

This orchid has a basal rosette of *oblong leaves,* marked with *round, blackish blotches.* The stout stem, with a few sheathing leaves, is topped by a loose spike of *purple flowers.* The upper flower-lobes and the lower lip are *folded back,* making the flowers look *squashed.* The plant grows in grassland, mountain ledges and open woodland, scattered around Scotland.

Plants
of the Coast

The Scottish mainland has around 3,910 kilometres of coastline, and the many islands add a further 6,290 kilometres. This constitutes two-thirds of the total UK coastline, so it is not surprising that Scotland has a rich variety of seaside flowers. The sea affects plants growing near the coast in three main ways: it warms the land, exposes the coasts to winds and salt-spray, and even floods over saltmarsh plants at the highest tides.

The warming influence of the sea on neighbouring land allows many cold-sensitive species to survive near the coastal strip. As a result, many flowers that are widespread in southern England become confined to coastal regions in Scotland.

Sand-dune systems are often particularly rich in flowers because their hummocks and hollows provide the shelter and sunny banks that encourage both flowers and pollinating insects, while the crushed shells in the sand produce calcium-rich soils in which many species flourish. Many grassland species flower here, along with more typical coastal plants such as Common Storksbill, Spring Vetch and Centaury. The damp hollows between the dunes support a variety of wetland plants, including Water Mint and Grass-of-Parnassus.

The relationship between plants and dunes is even more intimate, because it is plants which encourage the dunes to form. On the seaward side of dune systems, rapidly colonising plants like Sea Rocket trap wind-blown sand on the upper beach, allowing low hillocks to form. On these hummocks a tough species called Marram Grass (*Ammophila arenaria*) can establish, just beyond the reach of the

Oysterplant creeps and spreads to form bluish patches on shingly shores uncommonly around Scotland.

tides. Its broad leaves, which roll up into a tube in dry weather to reduce water-loss, act as an efficient trap for more sand. Marram has an extraordinary ability to grow rapidly upwards through the accumulating sand, until huge sand-hills develop.

The influence of the sea is not always so benign. Many Scottish coasts, especially in the west and north, are highly exposed, and strong winds and salt spray produce low, windswept grassland, in which only a few resilient plants like Spring Squill and Knotted Pearlwort can flower. In Caithness, Sutherland and Orkney, this clifftop grassland is the home of the rare Scots Primrose (p. 5), the only

Thrift or Sea-pink produces glorious displays of colour on rocky shores in early summer

PLANTS OF THE COAST

Saltmarshes support a range of highly adapted plants, such as the Sea Aster growing along this tidal creek on the Firth of Forth

instantly-recognisable Scottish endemic flower (that is, one found nowhere else in the world).

Plants of the cliffs include Sea Pink, Sea Campion, Roseroot and Scots Lovage. The name Scots Lovage is well-chosen, because it reaches no further south than the Mull of Galloway in the west and just over the English border to Lindisfarne in the east.

Perhaps the least visited of coastal habitats are saltmarshes. These form in sheltered estuaries and bays, where accumulated mud allows colonisation by a range of resistant plants, which are regularly inundated by the sea at the highest tides.

Lesser Meadow-rue
Thalictrum minus

BUTTERCUP FAMILY

Fl: 6-8 Ht: 15-40 cm (occasionally 1 m)

This easily overlooked perennial has *wiry, branching stems* with *leaves* divided into *many wedge-shaped lobes*. It has a *spreading inflorescence* of insignificant flowers with *4, greenish-yellow lobes* and a *hanging tassel of yellow stamens*. It grows in sand-dunes around the coast except for Shetland, with a larger form on lime-rich rocks in the S Highlands.

Roseroot
Sedum rosea

STONECROP FAMILY

Fl: 5-8 Ht: 15-30 cm

Roseroot grows on seacliffs, especially in the N and W, and on mountain rocks. It has *thick, flat, bluish leaves*, 1–4 cm long, and *rounded* heads of greenish-yellow, *4-petalled flowers*, 6mm across. The male flowers, with petals longer than sepals, grow on *separate plants* from the females, which have petals equalling the sepals and develop into *orange pods*.

Biting Stonecrop
Sedum acre
STONECROP FAMILY
Fl: 5-7 Ht: 2-10 cm

Resembling English Stonecrop (below), this is distinguished by its *yellow flowers*, about 12 mm across, and its *bright-green* (or occasionally reddish), fleshy, oval *leaves*, 3–5 mm long. Its creeping stems form succulent mats in sand-dunes, dry grassland, shingle and walls, inland in the S and E, round coasts elsewhere, but as a garden escape on the N Isles.

English Stonecrop
Sedum anglicum
STONECROP FAMILY
Fl: 6-9 Ht: 2-8 cm

A typical Stonecrop, this species forms *dense mats* of *flesh, balloon-like, reddish leaves*, about 4 mm long, from *creeping, rooting stems*. Its starry *flowers are about 12 mm across* with 5 *white petals, tinged pink beneath*. It is common on rocks, dry grassland and sand-dunes in W Scotland, but rarer in the E and Shetland, and absent from Orkney.

Purple Milk–vetch

Astragalus danicus

PEA FAMILY

Fl: 5-7 Ht: 5-35 cm

This creeping perennial is commonest in sand-dunes or coastal turf in the SE, Moray Firth area, W Isles and Galloway, but it also grows in lime-rich grassland inland in the E. It has *softly hairy leaves*, with *6–13 pairs* of narrow leaflets, ending in a leaflet, not a tendril. Its inflorescences, on leafless, *white-haired stalks* are roundish, *upward-facing clusters* of around 15 *blue-purple flowers*.

Kidney Vetch

Anthyllis vulneraria

PEA FAMILY

Fl: 6-9 Ht: up to 60cm

Kidney Vetch grows on sunny, lime-rich banks near the sea and in mountains, scattered around Scotland. Its *silky-haired* leaves, to 14 cm long, have *3–7 pairs of narrowly oval leaflets*, ending in a broader leaflet. It produces *roundish heads*, to 4 cm across, of *tightly-compressed pea-flowers* (p. 67) with *woolly sepal-tubes* and longer, *yellow or reddish petals*.

Spring Vetch
Vicia lathyroides

PEA FAMILY

Fl: 5-6 Ht: 5-20 cm

Like Common Vetch (p. 69) but with smaller flowers, and *no blotch underneath its leaf appendages*, this slender annual creeps over dune grassland or dry grassy banks around E and SW coasts. Its leaves have *2–4 pairs of narrow leaflets* and end in an *unbranched tendril* (often absent). The solitary, *blue-lilac flowers* are 6 mm long, and ripen into a *hairless pod*.

Whin
Ulex europaeus

PEA FAMILY

Fl: 12-6 Ht: 60-200 cm

'Officially' named Gorse, but generally called Whin in Scotland, this *evergreen shrub* is *armed with rigid, grooved spines* to 25 mm long. Its *golden-yellow, almond-scented pea-flowers*, 11–20 mm long, develop into *hairy, brown, explosive pods*. It grows in rough grassy places on acid soils, including clifftops, throughout Scotland, but is introduced on the Isles.

Burnet Rose
Rosa spinosissima
ROSE FAMILY
Fl: 5-7 Ht: 10-100 cm

Spreading by suckers, this shrub forms *bushy patches* in dunes, sandy heaths and inland limestone grassland round the mainland coast and inner W islands, but not on the W or N Isles. Its erect stems are densely covered in *straight prickles*, mixed with *stiff bristles*. It has leaves with *7–11 rounded, toothed lobes* and *creamy-white flowers*, 25–40 mm across.

Common Storksbill
Erodium cicutarium
GERANIUM FAMILY
Fl: 6-9 Ht: 2-15 cm

Related to Cranesbills (p. 24), Storksbill fruits have a *longer beak* (to 4 cm long). The only common Scottish species, this has a basal rosette of *feathery leaves*, *2–20 cm long*, and stout, hairy stems, topped by an *umbrella-like group* of up to 9 flowers, with *rosy-pink petals.* It grows in dune grassland and sandy fields near E and SW coasts and on the W Isles.

Sea Rocket
Cakile maritima
CABBAGE FAMILY
Fl: 6-8 Ht: 15-45 cm

Sea Rocket is a common strandline annual on sandy and shingly shores round most of Scotland and the islands. Its long tap-root produces patches of straggling stems with *fleshy, deeply-lobed, oblong* leaves. It has dense inflorescences of *mauve to white flowers*, about 15 mm across, with 4 petals, twice as long as the sepals, ripening into stubby, 2-jointed fruits.

Common Scurvy-grass
Cochlearia officinalis
CABBAGE FAMILY
Fl: 5-8 Ht: 5-50 cm

Three closely-related Scurvy-grasses grow on cliffs, saltmarshes and banks round Scottish coasts, and on salted roadsides inland. They are low-growing plants with white crucifer flowers, ripening to *globular, oval fruits*. This species has *heart-shaped basal leaves* and *white or mauve flowers*. Danish Scurvy-grass (*C. danica*) has *ivy-shaped leaves* and *pale lilac* flowers.

Thrift
Armeria maritima
THRIFT FAMILY
Fl: 4-10 Ht: 5-30 cm

Thrift grows on coastal rocks, pastures and saltmarshes round Scottish coasts, and, less commonly, on windy mountaintops. Its *narrow leaves* form *dense cushions*, from which arise *leafless flowerstalks*, topped by a *single, hemispherical head* of *rosy-pink to white* flowers, about *8 mm across*, with 5 petals, united at their base, and a funnel-shaped sepal-tube.

Sea Sandwort
Honckenya peploides
PINK FAMILY
Fl: 5-8 Ht: 5-25 cm

The leafy shoots of this *succulent* perennial form *dense patches* on sandy or shingly foreshores all round Scotland, as well as in W Isles machair. Its *stiff, yellow-green, pointed oval leaves* are *densely ranked* up stems which are topped by many *greenish-white flowers*, less than 1 cm across, with *5 narrow petals* as long as, or much shorter than, the sepals.

112

Knotted Pearlwort

Sagina nodosa

PINK FAMILY

Fl: 7-9 Ht: 5-25 cm

The only common Pearlwort with *relatively showy, 5-petalled flowers* (cf p. 37) to 1 cm across, this tufted perennial has *clusters of short, narrow leaves* which form 'knots' up its stem. It grows scattered around Scotland, on clifftops and other damp grassland, mostly near the coast. Sea Pearlwort (*S. maritima*) with *4-petalled* flowers, grows on dunes and seacliffs.

Greater Sea-spurrey

Spergularia media

PINK FAMILY

Fl: 6-9 Ht: up to 30 cm

Sea-spurreys resemble Corn Spurrey (p. 38) but have pink flowers and a *papery collar round the stem* below each leaf-whorl. This *hairless perennial* has *flowers 8–12 mm across*, with *pale pink or whitish* petals, *slightly longer* than the sepals. As a final check, a hand lens will show *broad wings* on its seeds. It grows in some saltmarshes around the coast and islands.

Sea Campion
Silene uniflora

PINK FAMILY

Fl: 6-8 Ht: 8-25 cm

Often abundant on sea-cliffs, coastal gravels and shingle around Scotland, Sea Campion has numerous *prostrate shoots* forming mats of *waxy, bluish, lance-shaped leaves* to 3 cm long. Its flowers, *20–25 mm across,* have an *inflated, bladder-like sepal-tube* with a *broad mouth,* and 5 *deeply notched petals* with a *prominent ruff of scales* around the flower's throat.

Babington's Orache
Atriplex glabriuscula

GOOSEFOOT FAMILY

Fl: 7-9 Ht: sprawling to 20 cm

The commonest Orache above the high-tide mark on sandy or gravelly shores round Scottish coasts, this is a sprawling, *mealy* annual. It has *thick, triangular, shallowly toothed leaves* and *slender, leafy spikes* of *tiny, greenish, unstalked flowers,* with male and female flowers on the same plant. The females develop into fruits enclosed by *2 diamond-shaped bracts.*

Glasswort

Salicornia europaea

GOOSEFOOT FAMILY

Fl: 8-9 Ht: 10-30 cm

Glassworts are a group of similar annuals, separated by details of their inflorescences and branching. They all have *succulent, cactus-like stems*, with paired leaves fused into *fleshy sheaths* around the stem, and minute *petal-less flowers* in *sheaths near branch ends*. They grow on open mud near the low tide mark of saltmarshes, scattered around Scottish coasts, particularly in the S and W.

Annual Seablite

Sueda maritima

GOOSEFOOT FAMILY

Fl: 8-10 Ht: 7-30 cm

A variable annual of saltmarshes and sea-shores, Annual Seablite grows scattered round Scottish coasts, usually below the high tide mark and often with Glasswort. It has red-tinged, *sprawling stems, fleshy, half-cylindrical leaves*, 3–25 mm long, and *tiny, greenish flowers*, with *2 minute bracteoles*, singly or in clumps of 2 or 3, *in the angles of upper leaves*.

Prickly Saltwort

Salsola kali

GOOSEFOOT FAMILY

Fl: 7-9 Ht: up to 60 cm

Another salt-tolerant member of the Goosefoot family (see previous page), this *sprawling, grey-green* annual grows on sandy shores around E and W coasts and islands, but not the W or N Isles. It has *short, fat, fleshy leaves*, ending in a *spiny tip*, and inconspicuous, greenish flowers in the angles of upper leaves, protected by *2 prickly bracteoles*.

Scots Primrose

Primula scotica

PRIMROSE FAMILY

Fl: 5-7 Ht: 1(!)-10 cm

This rare Scottish endemic (see p. 5) grows in windswept clifftop turf at just a few sites on the N coast and Orkney. It is far less conspicuous than visitors expect, with very short *leafless, mealy flower-stems* from a basal rosette of *spoon-shaped leaves* which are pale-green above and *whitish* beneath. Its *rich purple, tubular flowers* are about 8 mm across with a *yellow eye*.

PLANTS OF THE COAST

Sea Milkwort
Glaux maritima

PRIMROSE FAMILY

Fl: 6-8 Ht: 10-30 cm

This *rather fleshy, hairless* perennial with *oblong leaves* creeps amongst other vegetation in the upper zone of saltmarshes, or in salt-sprayed turf and rock crevices, round Scottish coasts and islands. Its *stalkless flowers*, in the angles of the upper leaves, lack petals, but they have a *delicate pink to white sepal-tube, 3–5 mm in diameter*, with *5 rounded lobes*.

Scarlet Pimpernel
Anagallis arvensis

PRIMROSE FAMILY

Fl: 6-8 Ht: 6-30 cm

Less common than in S Britain, this sprawling annual grows in dunes and light cultivated ground near E and SW coasts, on a few W islands, and as a rare garden weed on the N Isles. It has *weak, square stems*, with *stalkless, narrowly oval leaves*, and slender-stalked flowers, 12–15 mm across, with 5 spreading *scarlet or pink* (or rarely blue) petal-lobes.

Common Centaury
Centaurium erythraea
GENTIAN FAMILY
Fl: 6-10 Ht: 2-50 cm

This variable biennial has a basal rosette of *elliptical leaves*, 8–12 mm wide, and one or more erect, branching stems. The branches are topped by a *flat-topped cluster* of *stalkless, pink, tubular flowers*, with 5 spreading petal lobes about 1 cm in diameter. It grows in sand-dunes mostly round SE and W coasts and islands, with a few dry grassland sites inland.

Oysterplant
Mertensia maritima
BORAGE FAMILY
Fl: 6-8 Ht: spreading to 60 cm

Least rare on the N Isles, this plant of stony shores is scarce and declining in the S, with highly mobile populations everywhere, forming *bluish patches* on the shore. Its sprawling stems are less erect than shown (see p. 103), with opposite rows of *fleshy, oval, blue-washed leaves*. Its *tubular flowers,* in leafy inflorescences, are about 6 *mm across*. They *open pink then turn blue*.

Houndstongue
Cynoglossum officinale
BORAGE FAMILY

Fl: 6-8 Ht: 30-90 cm

This sturdy, *mousy-smelling* biennial grows in dunes and sandy ground, mainly in the SE. Its name comes from its *silkily hairy, tongue-shaped, grey leaves*. It has a few, stalked basal leaves and many stalkless stem-leaves which part-enclose the clusters of flowers. These have a *dull maroon* (rarely white) *petal-tube*, about 1 cm across, and ripen into *spiny nutlets*.

Sea Bindweed
Calystegia soldanella
BINDWEED FAMILY

Fl: 6-8 Ht: 10-60 cm

The main difference between this and other Bindweeds (p. 42) is that its stems *creep through the sand or shingle* of beaches, producing patches of *glossy, kidney-shaped leaves*. Amongst these appear solitary, *funnel-shaped flowers*, 3–5 cm across, with petals that are *pink with creamy stripes*. It is restricted to SW coasts, plus a few W islands and E coast sites.

Buckshorn Plantain

Plantago coronopus

PLANTAIN FAMILY

Fl: 5-7 Ht: up to 20 cm

Although its cylindrical inflorescence is unmistakably a plantain (see p. 86), this is the only Scottish species with *deeply-lobed leaves*. The *rather downy* plant often lies low over the ground, with leafless *flowerstalks curving upwards* from the middle of the leaf rosette and an inflorescence *up to 4 cm long*. It is common in bare ground and rocks near all our coasts.

Sea Plantain

Plantago maritima

PLANTAIN FAMILY

Fl: 6-8 Ht: up to 30 cm

This typical plantain has *tight rosettes of narrow, fleshy leaves* and *cylindrical flowerheads* with *conspicuous pale-yellow stamens*. It has *thicker, narrower, less strongly-veined leaves* than Ribwort Plantain (p. 87), and its flower-stalk is *not grooved*. It grows in saltmarshes and coastal turf around Scotland, and also beside mountain streams and marshes.

Henbane
Hyoscyamus niger

NIGHTSHADE FAMILY

Fl: 6-8 Ht: up to 80 cm

Now restricted in Scotland to sandy shores along SE coasts, this *foul-smelling* annual or biennial has a stout stem and *oblong, broadly-toothed leaves*, both covered in *sticky, white hairs*. It produces clusters of *urn-shaped, buff yellow flowers, veined with purple*, 2–3 cm across, with 5 petal-lobes. All parts of the plant, including the seeds, are very poisonous.

Carline Thistle
Carlina vulgaris

DAISY FAMILY

Fl: 7-10 Ht: 10-60 cm

Dried bracts form the 'everlasting flowers' of this thistle; the true florets are in the central disc. The *leaves are deeply-cut and prickly*, those of the 1st year making a *cottony rosette* which withers before the plant flowers. The *flowering stems are purplish and cottony*. It grows uncommonly in sand-dunes and lime-rich grassland in the E, SW and inner W islands.

Sea Aster
Aster tripolium
DAISY FAMILY
Fl: 7-10 Ht: 15-100 cm

Sea Aster is a *hairless perennial* with sturdy stems and *fleshy, lance-shaped, only slightly toothed leaves*. It has showy daisy flowers (p. 49), to 2 cm across, with *yellow disc florets* and spreading *blue-purple or pale lilac rays*. It is confined mainly to saltmarshes in sheltered estuaries and inlets, but also grows on salt-sprayed sea-cliffs and rocks, especially in the N.

Common Cornsalad
Valerianella locusta
VALERIAN FAMILY
Fl: 4-6 Ht: 5-40 cm

This *slender, hairless* annual grows most commonly in dune grassland, where its stems are often so short that the cup of *pale lilac, funnel-shaped flowers*, enclosed by bracts, sits tight against the sand. On walls or rocky outcrops, its brittle stems grow rather taller. It is found scattered through S Scotland and the W islands, more rarely in the N, and not on the N Isles.

Sea Holly
Eryngium maritimum

CARROT FAMILY

Fl: 7-8 Ht: 30-60 cm

Sea Holly has *fleshy, prickly, waxy, bluish-green leaves*, up to 12 cm across, with smaller, more lobed, stalkless stem leaves. It produces *egg-shaped heads*, about 2 cm across, of *steely blue flowers*, almost buried amongst spiny bracteoles. A plant of sandy and shingle foreshores, it grows uncommonly around the Solway and Clyde, and on a few W islands.

Scots Lovage
Ligusticum scoticum

CARROT FAMILY

Fl: 5-7 Ht: 15-90 cm

This tufted perennial grows on cliffs, coastal rocks and occasionally sand-dunes, all round Scotland and the Isles. It has *ribbed, reddish stems* and *glossy, hairless leaves* divided into 3 segments which are also 3-lobed. The typical carrot flowers (p. 57) are *greenish-white*, in heads to 6 cm across with 8–14 rays, and develop into *egg-shaped fruits with narrow wings*.

Wild Carrot
Daucus carota
CARROT FAMILY
Fl: 6-8 Ht: 30-100 cm

The wild ancestor of the cultivated carrot, this has *ridged, hairy stems* and *much-divided, feathery leaves*. It has a *flat-topped inflorescence* of *whitish flowers*, usually with a *deep purple central flower*, and *enlarged outermost petals*. Its fruits are *spiny, flattened* and *oval*. It grows on cliffs, dunes and rocky places near the sea in the W and more rarely in the E.

Sea Arrowgrass
Triglochin maritima
ARROWGRASS FAMILY
Fl: 7-9 Ht: 15-50 cm

This inconspicuous plant of saltmarshes and salt-sprayed turf all round the coast and islands has leafless stems and a basal tuft of *long, half-cylindrical, fleshy leaves*. It produces *dense spikes* of *tiny flowers* (top detail), *with 6 green lobes and a tufted white style*, ripening into *egg-shaped fruits* (lower detail). Marsh Arrowgrass (*T. palustris*) grows in upland marshes.

Eelgrass
Zostera marina
EELGRASS FAMILY
Fl: 6-9 Ht: up to 60 cm

Eelgrass grows submerged to depths of 4m below low water on muddy and sandy seabeds and in brackish lagoons around W coasts and islands, but is scarce and declining in E coast estuaries. Its *grass-like leaves* are 20–200 cm long and 5–10 mm wide, and its *minute flowers* are *partly enclosed in sheaths*, towards the base of leaf-like branches (see detail).

Spring Squill
Scilla verna
ASPARAGUS FAMILY
Fl: 4-6 Ht: 5-15 cm

This attractive perennial is common in coastal grassland and rocks along the N coast and in the N Isles, less common in SW Scotland and the W islands, and occasional on E coasts. Its *thick, glossy, narrow, rather twisted leaves* lie low amongst the grass, with leafless flower-stalks topped by short spikes of a few *violet-blue flowers* with *wide-spreading petal lobes*.

PLANTS OF LOWLAND WOODS

Although the Scottish Highlands are traditionally the main attraction for visitors, the lowlands also have much to offer. Indeed, anyone searching for wild flowers in springtime will find more colour and variety in lowland woods than in the Highlands, where the climate stops most plants from flowering until well into May,

Rather few natural woods remain in Scotland. In Argyll there are still some superb areas of oak woodland - at Glen Nant, near Taynuilt, or Taynish, south-west of Lochgilphead, for example. Gnarled and twisted by the winds, these woods are just as wild as the pinewoods of the central Highlands, and, in recognition of this, both have been declared as National Nature Reserves by Scottish Natural Heritage, the Scottish Government conservation body. Many of these woods were coppiced in the past to produce charcoal for the gunpowder industry in the area. Their main interest botanically is in the lichens and liverworts that carpet their trunks, but the floor beneath the trees is colourful in May and June with woodland flowers including Wood Anemone, Primrose and Wild Hyacinth.

In the central lowlands, almost the only natural woodland left after past felling clings to the steep banks of river gorges, notably along the Clyde near Lanark and in the valley of the Esk south of Edinburgh. Other important remnants of ancient woodland survive near Dalkeith in Midlothian and Hamilton in Lanarkshire, preserved as hunting parks.

Thanks to the enlightenment of the late eighteenth and early nineteenth century lairds

April is colourful in the woods of Argyll, with Primrose, Lesser Celandine and Common Dog-violet flowering beneath the trees

(landowners), fine planted woodland is now an attractive feature in many parts of the lowlands. These 'policy' woodlands (from the French for 'managed lands') were laid out to provide an attractive setting for the great houses of the lairds, as cover for game birds, as an investment in timber, and also, undoubtedly, for their scientific and amenity value. Non-native species such as Sycamore and Beech were often included in the planting, and it seems that the ground flora was often enhanced by attractive shrubs and herbs which would provide colour throughout the spring and summer.

Wild Hyacinth

Woodruff

As a result, these policy woods often host superb displays of non-native plants that are uncommon or absent elsewhere in Britain, including Leopardsbane, Pink Purslane, Pyrenean Valerian, and the winter-flowering White Butterbur - as well as natives such as Lesser Celandine, Wood Stitchwort, Sanicle and Ramsons.

Some of these aliens can confuse the unwary visitor. Few-flowered Garlic, for example, which flowers abundantly in May, is omitted from many field guides in which the nearest match is a rare Garlic from Cornwall! It spreads by tiny bulbils beneath its flowers, which break off when the plant is cut or grazed and grow into new plants, producing dense carpets in some lowland woods.

Wood Anemone

Anemone nemorosa

BUTTERCUP FAMILY

Fl: 3-5 Ht: 6-30 cm

In spring, Wood Anemone carpets deciduous woods and hedgerows throughout Scotland, except for the far N and W Isles. It is a rare introduction in the N Isles. It has *3, deeply-lobed stem leaves* in a whorl beneath its flowerhead, similar basal leaves, and a *single nodding flower*, 2–4 cm across, with *usually 6 or 7* (but up to 12) *white or pink-tinged petal-like lobes*.

Goldilocks Buttercup

Ranunculus auricomus

BUTTERCUP FAMILY

Fl: 4-5 Ht: 10-40 cm

This uncommon buttercup of nutrient-rich deciduous woodlands in S Scotland and the C Highlands is *less hairy* and *fewer flowered* than Meadow Buttercup (p. 18). It has *hairless, kidney-shaped lower leaves* and few stem leaves deeply divided into *narrow segments*. Its flowers, to 1 cm across, have *5 or fewer petals* which are soon shed and 5 *spreading sepals*.

Lesser Celandine
Ficaria verna
BUTTERCUP FAMILY
Fl: 2-5 Ht: 5-25 cm

A characteristic early flower of damp woodland, hedgerows and grassy banks throughout Scotland, apart from the wilder parts of the Highlands and NW, Lesser Celandine has *long-stalked, glossy, heart-shaped leaves* and *single flowers*, up to 3 cm across, with *8–12 narrow, glossy, yellow petals* and 3 *sepals which are shed early*. It can become a weed in gardens.

Wood Avens
Geum urbanum
ROSE FAMILY
Fl: 6-8 Ht: 20-60 cm

The stems of this *hairy perennial* arise from *a basal rosette of leaves with 2–3 pairs of toothed leaflets* and a much larger end leaflet. The flowers, *facing upwards* on long stalks, are 8–15 mm across, with 5 *spreading yellow petals* as long as the *narrow green sepals*. It grows in woods and shady places, but not in mountainous areas nor on the W or N Isles.

Wild Strawberry
Fragaria vesca
ROSE FAMILY
Fl: 4-7 Ht: 5-30 cm

Like Garden Strawberry, this wild form spreads by *rooting runners*, but its 'berries' are smaller and sweeter. Its leaves have *3 toothed leaflets, 1-6 cm long*, with *silky hairs pressed against the underside*. Its flowers are c. 15 mm across, with *5 overlapping white petals*. It grows in open woods and grassland all over Scotland except for some Highland areas.

Barren Strawberry
Potentilla sterilis
ROSE FAMILY
Fl: 3-5 Ht: 5-15 cm

Barren Strawberry resembles the previous species, but its fruit is a dry capsule. Its leaflets are *0.5–2.5 cm long*, with *spreading hairs on the underside* and *end in a tooth much smaller than its neighbours*. The flowers have *narrow petals, with the green sepals showing between them*. It grows in scrub and open woodland, except in the far N, W Isles and N Isles.

Wood Sorrel
Oxalis acetosella
WOOD-SORREL FAMILY
Fl: 4-8 Ht: 5-15 cm

This delicate perennial forms pockets of shining flowers from creeping underground stems on woodland floors or amongst shady mountain rocks throughout Scotland. Its leaves have *3, bright-green, wedge-shaped* leaflets, which *fold up* when not in sunshine. Its *solitary* flowers are about 1 cm across with *5 white or lilac petals, veined with deeper lilac.*

Dog's Mercury
Mercurialis perennis
SPURGE FAMILY
Fl: 3-5 Ht: 15-40 cm

This creeping perennial carpets the floors of well-established woods, almost entirely S of the Great Glen. Its *downy* stems have *opposite pairs of oval, toothed leaves.* Its flowers are *4 mm across, with 3 green lobes,* in spikes in the angles of upper leaves. Each patch of plants has either male flowers with yellow stamens (M on right), or shorter female flowers (F).

Tutsan

Hypericum androsaemum

ST. JOHN'S-WORT FAMILY

Fl: 6-8 Ht: 40-100 cm

This *semi-evergreen shrub* grows in damp woods and hedges in the far W, from Galloway to Ross-shire and the inner islands, with a few introductions elsewhere. Its stem has many *large, unstalked, oval leaves*. Its *flowers*, which grow *in small clusters*, are *about 2 cm across*, with *5 blunt petals* and *5 unequally-sized sepals*, ripening to purplish-black berries.

Common Enchanter's Nightshade

Circaea lutetiana

WILLOWHERB FAMILY

Fl: 6-8 Ht: 20-70 cm

Widespread in shady woodland in the S and W, but rare in the N, this perennial has *oval, toothed, stiffly hairy* leaves and an *open spike* of flowers, 4–7 mm across, with *2 deeply-notched white petals* and *2 pale green sepals*. Alpine Enchanter's Nightshade (*C. alpina*), with *hairless leaves* is commoner in the north, as is the sterile hybrid between the two.

Three-nerved Sandwort

Moehringia trinervia

PINK FAMILY

Fl: 5-6 Ht: 10-40 cm

Found in rich woodland soils through the lowlands and inner W islands, this *straggling annual* resembles Common Chickweed (p.36) but its stem is *downy all round* and its *egg-shaped leaves are strongly 3–5 veined underneath*. Its flowers are around 6 mm across, with *undivided petals about half as long as the narrow sepals* which *have broad white margins*.

Wood Stitchwort

Stellaria nemorum

PINK FAMILY

Fl: 5-6 Ht: 15-60 cm

Stitchworts (see also p. 136) are fragile herbs with narrow leaves and *starry white flowers* with *5 sepals, 5 notched petals and 10 stamens*. Wood Stitchwort grows in damp woods and shady streamsides in the lowlands. Its *broadly oval leaves* end in a *short point*. Its flowers, 10–18 mm across, have *petals divided almost to the base* and *twice as long as the sepals*.

Greater Stitchwort
Stellaria holostea
PINK FAMILY
Fl: 4-6 Ht: 15-60cm

This Stitchwort (see p. 135) grows in woods and hedges on the mainland and inner islands, but not the W Isles and introduced to the N Isles. More robust than Lesser Stitchwort (below), it has *squarish stems* and *narrow, rough-edged leaves*, 4–8 cm long, *tapering to a long point*. Its flowers, with *petals longer than the sepals*, are *notched to about half-way down*.

Lesser Stitchwort
Stellaria graminea
PINK FAMILY
Fl: 5-8 Ht: 20-90 cm

Weaker and slenderer than Greater Stitchwort, this species has *greener leaves, smooth at the margins* and *without a drawn-out point, smoothly 4-angled stems* and *smaller flowers* (to *12 mm across*), developing later in summer. It is more widespread in woods and open heathland on the mainland, scattered in the NW and W Isles, and introduced to Shetland.

Spring Beauty
Claytonia perfoliata

BLINKS FAMILY

Fl: 4-7 Ht: 10-30 cm

This pale-green annual has fleshy, diamond-shaped lower leaves and a single pair of stem leaves. These are united into a ruff around the stem beneath a small group of white flowers, 5–8 mm across, with 5 unnotched petals. The species was introduced from North America around 1852, and grows in poor sandy soils at a few sites in the E and SW and on Shetland.

Pink Purslane
Claytonia sibirica

BLINKS FAMILY

Fl: 4-7 Ht: 14-40 cm

Introduced around 1838 from N America, this hairless plant is common in woodland through S, C and E Scotland, and in plantations on the W and N Isles. It has *long-stalked, fleshy basal leaves* and *a single pair of fleshy, stalkless stem leaves*. Its flowers, *15–20 mm across*, have *2 sepals* and *5 longer, deeply notched pink or white petals, with darker pink veins*.

Primrose
Primula vulgaris
PRIMROSE FAMILY
Fl: 2-6 Ht: 5-12 cm

Found throughout Scotland in woods and open, grassy places, this familiar plant has basal rosettes of *crinkly, oblong, blunt-ended leaves*, to 25 cm long, which *narrow gradually into a stalkless base* (cf Cowslip, p. 80). Its flowers have a *yellow petal-tube with 5 spreading, shallowly-notched lobes, about 3 cm in diameter*, surrounded by a *shaggy sepal-tube*.

Yellow Pimpernel
Lysimachia nemorum
PRIMROSE FAMILY
Fl: 5-9 Ht: up to 40 cm

This perennial grows in shady woods and hedgerows over most of Scotland, although rare in Orkney and the W Isles, and absent from Shetland. It has *opposite pairs of evergreen, oval, pointed leaves* and solitary *tubular, pale yellow flowers with a deeper yellow 'eye'*, on *fine stalks just longer than the leaves*. Its petal lobes *spread out* to a diameter of about 12 mm.

138

Creeping Jenny
Lysimachia nummularia

PRIMROSE FAMILY

Fl: 6-8 Ht: up to 60 cm

Reminiscent of Yellow Pimpernel, this creeping perennial has leaves that are *almost circular in outline, blunt-ended and dotted with black glands*. Its flowers are *more cup-shaped, richer yellow, gland-dotted, 15–25 mm across*, and borne on *stout stalks shorter than the leaves*. Non-native in Scotland, it grows in damp hedgebanks and grasslands, mainly in the S and E.

Woodruff
Galium odoratum

BEDSTRAW FAMILY

Fl: 5-6 Ht: 15-45 cm

Woodruff spreads by creeping runners to form *pale-green patches* in damp woodlands, in all but Highland areas and the Isles. It has unbranched, *4-angled stems*, with *whorls up the stem of 6–9 narrow leaves* with *forwardly-directed marginal prickles*. Its flowers, in loose clusters, are *fragrant, white, tubular, with 4 narrow petal-lobes*, and about 4 mm across.

Lesser Periwinkle
Vinca minor
PERIWINKLE FAMILY
Fl: 3-5 Ht: 30-60 cm

A garden escape from mainland Europe, this *evergreen shrub* has *trailing stems* which *root at intervals* and give off erect flowering stems. It has *opposite pairs of short-stalked, glossy, oval, privet-like leaves* and solitary *blue flowers*. These are *25–30 mm across, with a long tube and 5 spreading lobes*. It grows in scattered woods and hedgerows, mostly in the S and E.

Green Alkanet
Pentaglottis sempervirens
BORAGE FAMILY
Fl: 5-6 Ht: 30-100 cm

Introduced from SW Europe, this *roughly hairy perennial* grows in hedgerows and woodland margins locally in the S and E. It has *oval, pointed, net-veined leaves*, somewhat resembling Comfrey (p. 172) but with *bright blue, tubular flowers* with *5 sepal-teeth* and *5 petal-lobes*. These are similar to Bugloss flowers (p.41), but have a *straight, not kinked, petal-tube*.

Wood Forget-me-not
Myosotis sylvatica
BORAGE FAMILY
Fl: 5-8 Ht: 15-45 cm

This *downy* biennial to perennial grows in damp woods in E and SW Scotland and occasionally as a garden escape elsewhere. It is distinguished from other Forget-me-nots by its habitat, the *relative short flower stalks (about 1½ times as long as the sepal-tube)*, the *hooked hairs covering the sepal-tube* and the large *sky-blue flowers, 6–8 mm in diameter*.

Yellow Figwort
Scrophularia vernalis
FIGWORT FAMILY
Fl: 4-6 Ht: 30-80 cm

Typical figworts (see p. 142) have *square stems* and leafy clusters of flowers with globular, helmeted, rather dingy-coloured petal-tubes. However, this species, introduced to a few plantations in the SE and Moray Firth area, has *weakly-angled*, hairy stems, *thin, wrinkled leaves* and *greenish-yellow* flowers *without an obvious helmet* but with a *contracted throat*.

Common Figwort
Scrophularia nodosa
FIGWORT FAMILY
Fl: 6-9 Ht: 40-80 cm

This typical Figwort (p. 141) grows in damp woods and hedgerows throughout the mainland and W islands. Its stout stem is *sharply angled but not winged* and its *foul-smelling, pointed, oval* leaves have *coarsely-toothed margins* and *unwinged* stalks. Its flowers have a *greenish petal-tube* with a *purplish-brown helmet* and their 5 sepal-lobes have a *narrow, pale border*.

Common Cow-wheat
Melampyrum pratense
BROOMRAPE FAMILY
Fl: 5-10 Ht: 8-60 cm

The roots of Cow-wheat tap those of other plants to draw nutrients. It has *opposite pairs of narrow, short-stalked leaves* and *tubular yellow flowers held horizontally in pairs* and *turned to face the same way*. The *11–15 mm long petal-tube* is twice as long as the sepal-tube, with a *red-tinged upper lip*. It grows in open woods, heaths and grassland throughout Scotland.

Toothwort
Lathraea squamaria
BROOMRAPE FAMILY
Fl: 4-6 Ht: 8-30 cm

Toothwort has no green, photosynthetic leaves because it lives as a parasite on various trees. It has *stout, downy, flesh-coloured stems*, topped by a *one-sided cluster* of *pinkish, 2-lipped, tubular flowers*. It grows in a few broadleaved woods in the S (cf. Bird's-nest Orchid, p. 150). Stouter, redder Broomrapes (*Orobanche* spp) parasitize Thyme and Whin in the W.

Giant Bellflower
Campanula latifolia
BELLFLOWER FAMILY
Fl: 7-8 Ht: 50-120 cm

The tallest Bellflower, this hairy perennial grows in rich woodland and hedgerows, scattered through the lowlands. It has *bluntly-angled, unbranched stems*, and *toothed, egg-shaped leaves*, the lower with a *winged stalk* and the upper *stalkless*. The *purplish-blue, bell-shaped flowers* are *4 cm or more long*, with *petal-lobes almost as long as the tube at their base*.

Goldenrod
Solidago virgaurea
DAISY FAMILY
Fl: 7-9 Ht: 5-75 cm

This native perennial of dry woodland and grassland occurs throughout Scotland, although commoner in the W. It has *slightly-toothed, stalked, lower leaves and narrower, unstalked upper leaves*. Its flowers, in *showy spikes*, are *golden-yellow*, up to 1 cm across, *with 6–12 spreading rays* and surrounded by *several rows of overlapping, greenish-yellow bracts*.

Leopardsbane
Doronicum pardalianches
DAISY FAMILY
Fl: 5-7 Ht: 30-90 cm

This species from W Europe, was widely planted in policy woodlands (see p. 128) for its splashes of *yellow flowers, about 5 cm across*. Dense patches still survive in woods today, widely in the E and less commonly in the W. It has *basal rosettes of long-stalked, rounded, toothed leaves* and upper leaves which *clasp and half-enfold the stem at their base*.

White Butterbur
Petasites albus
DAISY FAMILY
Fl: 2-5 Ht: 10-50 cm

Originally introduced from C Europe for a flush of spring colour in woodlands, this patch-forming perennial has become a troublesome weed in woods in C and E Scotland. In early spring, it produces *pyramidal spikes of dirty-white tubular flowers* with no ray-florets. These are followed by patches of *shallowly-lobed leaves*, about 20 cm across (cf Butterbur, p. 180).

Moschatel
Adoxa moschatellina
MOSCHATEL FAMILY
Fl: 4-6 Ht: 5-10 cm

Moschatel grows in rich woodland throughout the S and E and amongst shady mountain rocks. Its *long-stalked root leaves* are *divided into three 3-lobed leaflets*. The stem has a *single, opposite pair of short-stalked, 3-lobed leaves*, topped by a head of 4 greenish flowers, *with 5 petal-lobes, pointing at right angles and a 5th flower, with 4 lobes, pointing upwards*.

Honeysuckle
Lonicera periclymenum

HONEYSUCKLE FAMILY

Fl: 6-9 Ht: climbing to 6 m

Honeysuckle twines round tree trunks or scrambles amongst rocks and hedgerows almost anywhere in Scotland. It has *woody stems* and *opposite pairs of untoothed, oval leaves*. The stems are topped by *clusters of 2-lipped flowers*, with *petal-tubes about 4 cm long, creamy-white inside and purplish or yellowish outside*. These develop into *glossy red berries*.

Pyrenean Valerian
Valeriana pyrenaica

VALERIAN FAMILY

Fl: 6-7 Ht: up to 1 m

An introduction from the Pyrenees around the 17th century, this perennial is now well established in damp woodland and shady riversides in C and E Scotland as far N as the Moray Firth. It is taller than Common Valerian (p. 97), with *slightly larger, denser flower-heads, broader, undivided lower leaves* and *upper stem-leaves with 1 or 2 pairs of leaflets*.

Common Ivy
Hedera helix
IVY FAMILY
Fl: 9-11 Ht: climbing to 30 m

The woody stems of Ivy form carpets on woodland floors or climb in trees, hedges or rocks almost throughout Scotland, except Shetland. Its *dark-green, hairless leaves* have *3 or 5 spreading lobes*, with upper stem leaves unlobed. It produces *umbrella-like clusters of greenish-yellow, 5-petalled flowers*, about 5 mm across, which develop into *black berries*.

Sanicle
Sanicula europaea
CARROT FAMILY
Fl: 5-8 Ht: 20-60 cm

Characteristically growing in dappled sunlight on the floor of deciduous woods throughout lowland areas and the inner islands, but not on the W or N Isles, this *hairless* perennial has thin, *long-stalked, glossy leaves, divided into 5 toothed lobes* and a *branching inflorescence of pink or white flowers, in ball-like clusters*, which ripen into *bristly, egg-shaped fruits*.

Pignut
Conopodium majus
CARROT FAMILY
Fl: 5-7 Ht: 30-50 cm

The leaves of Pignut are triangular in outline and finely divided into narrow segments. The basal leaves wither as the flowers develop. The white flowers are arranged in a neatly convex umbel (see p. 57), which droops in bud and usually has no bracts. It grows in open woodland and grassy places on acid soils throughout Scotland but is introduced on Shetland.

Lords-and-Ladies
Arum maculatum
ARUM FAMILY
Fl: 7-8 Ht: 30-50 cm

This patch-forming perennial has *glossy, dark-green, arrow-shaped* root-leaves, which appear in early spring. Later, a *pale, yellow-green or purple-edged sheath* develops which *completely encloses* the *purple-tipped flowering spike*. The flowers eventually ripen into red poisonous berries. The plant grows in rich woodland in S, C and E Scotland but is probably not native.

Herb Paris
Paris quadrifolia
HERB PARIS FAMILY
Fl: 5-8 Ht: 15-40 cm

From a creeping underground stem, this perennial produces hairless aerial stems with a *ring of usually 4, broad, unstalked, oval leaves*. A single flower arises from the middle of these, with *4 narrow green sepals, 4 shorter and narrower yellowish-green petals and 8 stamens*. It grows uncommonly in a few lime-rich woods, mostly in C Scotland and the C Highlands.

Broad–leaved Helleborine
Epipactis helleborine
ORCHID FAMILY
Fl: 7-10 Ht: 25-80 cm

This robust orchid grows in woods, clearings and a few old gardens in S and C Scotland and the far W. Its stems are *white-downy above* with a spiral of *pointed, broadly oval leaves half-clasping the stem*. It produces a *crowded, one-sided spike of 15–50 flowers* with *3 narrow, green or dull-purple outer lobes, 2 pinkish upper lobes* and a *purple, heart-shaped lower lip*.

Bird's-nest Orchid
Neottia nidus-avis

ORCHID FAMILY

Fl: 6-7 Ht: 20-45 cm

Entirely lacking green leaves, this honey-coloured orchid relies on a fungus in its roots to absorb nourishment from the leaf-litter of shady woods in C Scotland and the C Highlands. Its stems are enwrapped by a few *papery, brown, sheathing leaves* and topped by a spike of *sickly-smelling brown flowers* with a *lower lip splitting into 2 blunt-ended lobes.* (cf p. 143).

Few-flowered Garlic
Allium paradoxum

ONION FAMILY

Fl: 4-6 Ht: 20-50 cm

This introduction forms dense patches in a few woods in the S and Moray coast. The single leaf from each bulb is *narrower (5–25 mm wide), glossier and darker green* than that of Ramsons (opposite). The inflorescence, on a *strongly 3-sided stem*, has *1–4, long-stalked, starry white flowers*, intermixed with *many yellow bulbils* and protected by a *papery bract*.

Ramsons
Allium ursinum
ONION FAMILY
Fl: 4-6 Ht: 10-45 cm

Ramsons or Wild Garlic forms carpets of *bright green, tongue-shaped leaves*, 10–25 cm long and up to 8 cm broad, *smelling strongly of garlic when crushed*. Its *leafless, angled* flower-stem is topped by a *flattened cluster* of *6–20 long-stalked, starry white flowers*, with a 2-lobed papery bract beneath. It grows in damp woods and shady places, mostly in the S and W.

Snowdrop
Galanthus nivalis
ONION FAMILY
Fl: 1-4 Ht: 15-25 cm

Widely planted in woods and grassy banks throughout lowland Scotland, this delicate perennial has *strap-shaped leaves* about as long as the leafless flower-stalk which is topped by a solitary, nodding, bell-shaped flower. This has *3 pure-white outer flower-lobes*, about 15 mm long, and *3 shorter inner flower-lobes* which are *white with a green spot at their tip*.

Lily-of-the-Valley
Convallaria majalis
ASPARAGUS FAMILY

Fl: 5-6 Ht: up to 35 cm

This fragrant plant is native in a few lime-rich woodlands in the C Highlands, but introduced elsewhere, especially in C Scotland. It forms *dense patches* of *broadly oval root leaves*, 8–20 cm long. Their twisted bases sheath a leafless flower-stem, topped by a *one-sided spike of 6–12, nodding, bell-like white flowers*, 5–9 mm long, which develop into *round, red berries*.

Solomon's-seal
Polygonatum multiflorum
ASPARAGUS FAMILY

Fl: 5-6 Ht: 30-80 cm

This showy garden plant is a long-established introduction in a few woods in the SW and E. It grows as open patches of *hairless, arching stems*, with a *few, stalkless, oval stem leaves*. In the angles of upper leaves it produces *hanging tassels of 1–6, nodding, tubular, greenish-white flowers*, 9–15 mm long, with *constricted waists*. These develop into *blue-black berries*.

Wild Hyacinth
Hyacinthoides non-scripta
ASPARAGUS FAMILY
Fl: 4-6 Ht: 20-50 cm

Wild Hyacinth or Bluebell grows as scattered patches of *strap-shaped leaves* in woods and shady gullies around Scotland, except in Highland areas and the N Isles, and forms carpets in a few woods in the S. Its flower-stalks are topped by a *one-sided spike of 4–16, nodding, violet-blue (rarely white or pink) bell-shaped flowers*, with petal-lobes bent back at the tip.

Great Woodrush
Luzula sylvatica
RUSH FAMILY
Fl: 5-6 Ht: 30-80 cm

This robust plant forms *tussocks of bright green, strap-shaped leaves*, up to 30 cm long by about 1 cm wide. From these arise flowering stems topped by a *loosely-branched cluster of small, chestnut-brown flowers* in groups of 3 or 4, which ripen to brown egg-shaped fruits. The plant grows in acid woodland, gullies, mountain rocks and clifftops throughout Scotland.

PLANTS OF RIVERS, PONDS AND MARSHES

In their steep journey from the mountains, many of Scotland's rivers are fast flowing, and so provide limited root-hold for water plants. On their humid banks a variety of water-loving species grow, including the large golden-yellow flowers of Marsh Marigold, the pink heads of Cuckooflower or the frothy blossoms of Meadowsweet. Where the rivers widen and slow as they approach the coast, more plants can become established in the water: beautiful, drifting rafts of Water Crowfoots, for example, perhaps along with Water-milfoil and Marestail. Byways in Scotland's canals can be even richer in water plants.

Like lowland woods (p. 126), Scotland's riverbanks have their share of non-native species, some of which have become so abundant in places that they displace native species. Indian Balsam, a native of the Himalayas, is widespread along the Water of Leith in Edinburgh or the Clyde at Dalmarnock in Glasgow, spread by waterborne seeds from its explosive fruits to form dense hedges in places.

A more pernicious alien is Giant Hogweed from the Caucasus. Victorians introduced it into their gardens to celebrate its mammoth size (up to 5 m tall), but it soon escaped, spreading along railways and riverbanks. Today it is abundant along parts of the Rivers Clyde and Kelvin in Glasgow, the Tweed and Teviot in the Borders, and rivers around Edinburgh. Because its sap can sensitise the skin to sunlight, causing severe sunburn, attempts are being made to eradicate it in many areas.

After several decades of neglect and infilling, ponds are again being regarded as an important

Giant Hogweed is an alien from south-west Asia which has established dense thickets along riverbanks in the lowlands

feature of lowland landscapes, especially on farms, and new ponds have been dug into which the natural flora soon moves, often on the feet of birds. The tiny, floating Duckweeds are usually first to arrive, but soon other species such as Water Starwort, Water Crowfoot, Amphibious Bistort and Pondweeds will follow. Many new ponds are now maturing attractively, although, in the absence of disturbance by drinking cattle and horses, they will need regular dredging to maintain their diversity.

Many of these wetland species also grow around the edges of the lochs dotted around the lowlands,

Blood-drop-emlets (Mimulus luteus) is one of several cultivated species and hybrids of Monkeyflower which have escaped and now grow on riverbanks and in ditches around Scotland.

*Pick-a-back-plant (*Tolmeia menziesii) *is a member of the Saxifrage family from North America which has become established in woods and along rivers in southern Scotland*

but the larger size and greater depth of these water bodies also allows dense patches of Reeds (*Phragmites australis*) or Bottle Sedge (*Carex rostrata*) to grow out into the open water. Amongst the reeds, in damp woods around the loch edge, and in overgrown marshy areas can be found plants like Meadowsweet, Purple Loosestrife, Cowbane, Hemlock and Hemlock Water-dropwort, as well as Yellow Loosestrife and rarer relatives such as Tufted Loosestrife (*Lysimachia thyrsiflora*).

Many of the fine Highland lochs are acidic in nature because of the natural acids that wash from surrounding peatlands, and these tend to support a less diverse flora (see p. 203).

Marsh Marigold
Caltha palustris
BUTTERCUP FAMILY

Fl: 3-7 Ht: up to 60 cm

This hairless perennial forms splashes of yellow in spring in marshes, ditches, streamsides and wet woods throughout Scotland. Its *glossy, kidney-shaped leaves* have a *wavy, toothed edge*. The flowers, which are *up to 5 cm across*, have *5–8 golden-yellow, petal-like lobes*, often greenish beneath, and *up to 100 stamens*, maturing into clusters of *dry, pod-like fruits*.

Ivy-leaved Water Crowfoot
Ranunculus hederaceus
BUTTERCUP FAMILY

Fl: 6-9 Ht: creeping 10-40 cm

Water Crowfoots are very variable *water-living, white-flowered buttercups* with *rounded floating leaves* or *feathery underwater leaves* or both. This is the commonest Scottish species, growing in wet mud and shallow water round most of the country. It has *smallish flowers, to 8 mm in diameter, fleshy, ivy-shaped floating leaves* and *no submerged leaves*.

Common Water Crowfoot

Ranunculus aquatilis

BUTTERCUP FAMILY

Fl: 5-6 Ht: creeping to 1 m

Frequent in the shallow water of ponds, ditches and slow-moving streams in the E and SW, and at scattered sites in the W and on W islands and Orkney, this species has floating leaves with 3–7 deeply-cut, wedge-shaped, toothed lobes, feathery submerged leaves and flowers to 18 mm across. It seems to benefit from some disturbance by farm animals etc.

Round-leaved Water Crowfoot (*R. omiophyllus*), like Ivy-leaved (opposite) but with *flowers to 15 mm across* and *roundish to kidney-shaped floating leaves*, grows in streams and mud in the SW.

River Water Crowfoot (*R. fluitans*) with *only submerged, much-forked leaves to 50 cm long*, and *flowers to 25 mm across*, grows in fast-flowing E streams.

Five other species occur rarely, including one (*R. baudotii*) in brackish coastal pools.

Lesser Spearwort
Ranunculus flammula
BUTTERCUP FAMILY
Fl: 5-9 Ht: creeping to 80 cm

This abundant but very variable plant grows in wet places throughout Scotland. It has *small buttercup flowers*, to 2 cm across, on *grooved and slightly hairy stalks*, with *5 greenish-yellow sepals* and *5 pale-yellow petals*. Its *stems creep* amongst the vegetation, with shoots bending upwards and *rather fleshy, lance-shaped leaves*, the upper of which are stalkless.

Celery-leaved Buttercup
Ranunculus sceleratus
BUTTERCUP FAMILY
Fl: 5-9 Ht: 20-60 cm

This annual herb has *shiny, toothed, deeply-lobed* root leaves and *less divided* stem leaves. Its buttercup flowers, less than 1 cm across, have pale-yellow petals and bent-back sepals, producing heads of 70–100 tiny green fruits. It grows in and around muddy ponds, streams and ditches, in S and C Scotland and coastal areas N to the Moray Firth and Caithness.

Opposite-leaved Golden Saxifrage

Chrysosplenium oppositifolium

SAXIFRAGE FAMILY

Fl: 4-7 Ht: 5-15 cm

This perennial of springs, stream-sides and wet ground throughout Scotland, except Shetland, forms *loose mats of creeping, rooting stems* with *opposite pairs of bluntly toothed, rounded, pale-green, sparsely-hairy leaves*. Its flowers, 3–4 mm across, have *no petals* but *4 or 5 greenish-yellow sepals* surrounded by a *cup-like frill of bright green bracts*.

Alternate Water-milfoil

Myriophyllum alterniflorum

WATER-MILFOIL FAMILY

Fl: 5-8 Ht: 20-120 cm

The spikes of Water-milfoil project from lochs, streams and ditches, most commonly in the NW and Isles. Its slender, branching shoots have *whorls of 4 (sometimes 3) leaves* with *6–18 feathery segments*. Tiny flowers grow in groups of 1–4 in the angles of leafy bracts. Spiked Water-milfoil (*M. spicatum*) with whorls of 4 leaves with *13–38 segments* is less common.

Meadowsweet
Filipendula ulmaria
ROSE FAMILY
Fl: 5-9 Ht: 60-120 cm

The *frothy, cream-coloured inflorescences* of this perennial appear abundantly in marshland, wet meadows and riverbanks throughout Scotland. Its leaves are divided into *5–11 stalked, toothed, oval leaflets, often silver-hairy underneath*. Its flowers, *5–10 mm across with 5 sepals and 5 (or occasionally 6) petals*, develop into *spirally-twisted heads of 6–10 dry fruits*.

Marsh Cinquefoil
Comarum palustre
ROSE FAMILY
Fl: 5-7 Ht: 15-45 cm

The underground stems of this perennial creep through marshes, bogs and wet moorland round most of Scotland. Its long-stalked leaves, often bluish-green underneath, have 3, 5 or 7 oblong, toothed leaflets, to 6 cm long, spreading like fingers. Its flowers, in an open inflorescence, have 5 deep-purple petals, 5 longer, purple sepals and 5 narrow, green bract-lobes.

Water Avens
Geum rivale
ROSE FAMILY

Fl: 5-9 Ht: 20-60 cm

Water Avens grows in marshes, riversides and damp ground throughout Scotland, but not the W Isles. Its rosette leaves have *3–6 pairs of oval, toothed side leaflets* and a *larger end leaflet*. The stem leaves are *undivided or 3-lobed*. Its *nodding, lantern-shaped* flowers, to 15 mm long, have 5 *notched, peach-coloured* petals almost concealed by a *purple sepal-tube*.

Grass–of–Parnassus
Parnassia palustris
GRASS-OF-PARNASSUS FAMILY

Fl: 7-10 Ht: 10-30 cm

This perennial produces an *erect tuft of long-stalked, pale-green, heart-shaped root-leaves*. A *single, stalkless leaf* clasps its flower-stem. The *flowers are white, veined* and 15–30 mm across, with 5 sepals, 5 petals and 5 stamens alternating with 5 *frilly tufts* of sterile stamens. It grows in base-rich wet grassland and marshes throughout Scotland, except the W Isles.

Purple Loosestrife
Lythrum salicaria
LOOSESTRIFE FAMILY
Fl: 6-8 Ht: 60-120 cm

Purple Loosestrife has *narrow, pointed oval, downy* leaves. Its stems are topped by *dense spikes*, to 30 cm long, of flowers that are 10–15 mm across with *greenish-purple, toothed sepal-tubes*, usually *6 narrow, purple petals* and 12 stamens. It inhabits marshes, riverbanks and lochsides, mostly in the W from Skye southwards, with scattered sites elsewhere.

Water Purslane
Lythrum portula
LOOSESTRIFE FAMILY
Fl: 6-10 Ht: 4-25 cm

This *hairless annual* has *4-angled, reddish, rooting stems* which *creep* through muddy pool edges and wet, open ground in scattered sites across Scotland, except for the NW and Shetland. It has *minute green flowers*, tucked in the angles of its opposite pairs of *fleshy, spoon-shaped* leaves. These have *6 petals which soon drop* and a *6-toothed sepal-tube*.

WILLOWHERBS

Willowherbs (*Epilobium* species) (e.g. Hoary Willowherb, right) are perennials with narrow, toothed leaves and pinkish flowers with 4 petals, 8 stamens and a *club-shaped or 4-lobed stigma*. Their narrow fruit capsules split lengthwise to release white-plumed seeds. They differ from Rosebay Willowherb (p.23) in having *upright flowers* with *equally-sized petals*.

Great Willowherb
Epilobium hirsutum
WILLOWHERB FAMILY
Fl: 7-8 Ht: 80-150 cm

The stem of this species is downy with spreading white hairs and is half-clasped by pairs of *hairy, lance-shaped leaves*. Its pinkish-purple flowers, 15–25 mm across, have *shallowly-notched petals* and a creamy stigma. It grows by streams and marshes in the S and E. Hoary Willowherb (*E. parviflorum*) (see box), with *unpaired, non-clasping* leaves is less common in similar habitats.

Short–fruited Willowherb

Epilobium obscurum

WILLOWHERB FAMILY

Fl: 7-8 Ht: 30-60 cm

This willowherb grows in marshes, streamsides and wet woods throughout Scotland, and as a recent colonist on Shetland. Its *downy stems* are *marked with 4 raised lines* running down from the *stalkless base* of its *spear-shaped leaves*. Its *purplish-pink flowers* are 6–8 mm across with *shallowly-notched petals*, a *club-shaped stigma*, and a *sticky-haired sepal-tube*.

Marsh Willowherb

Epilobium palustre

WILLOWHERB FAMILY

Fl: 7-8 Ht: 15-60 cm

Marsh Willowherb spreads by thin runners in marshes and bogs all round Scotland. It has *downy stems* and *unstalked, lance-shaped leaves*, to 7 cm long, mostly in *opposite pairs*. Its flowers, *held almost horizontally*, are 4–6 mm across with *pale rose or white, shallowly-notched petals* and a *club-shaped stigma*, ripening to a *downy capsule*, 5–8 cm long.

Marsh Yellow-cress
Rorippa palustris

CABBAGE FAMILY

Fl: 6-9 Ht: 8-60 cm

This *hairless annual*, with *hollow, branching stems*, has leaves *divided into 5–13 wavy-edged lobes* but *not cut to the midrib*, often with *small ears clasping the stem* at their base. Its flowers, in *loose heads*, are *3 mm across* with *4 pale-yellow petals*, ripening into an *oval pod, 5–10 mm long*. It grows in damp muddy places, in the S with a few sites in the NE.

Watercress
Nasturtium officinale

CABBAGE FAMILY

Fl: 5-10 Ht: 10-60 cm

Watercress is native in streams, marshes and ponds around Scotland. It has *creeping, rooting, dark-green to purple stems* and *hollow flower-stems*, which bend upwards or float on the water. Its leaves have *2–10 oval side leaflets, and a larger end leaflet*. Its tight clusters of *white 4-petalled flowers*, around *6 mm across*, ripen into *slightly curved pods*.

Cuckooflower
Cardamine pratensis
CABBAGE FAMILY
Fl: 4-6 Ht: 15-60 cm

Also known as Lady's Smock, this graceful perennial grows in damp grassland, marshes, and ditches all over Scotland. It resembles a large Wavy Bitter-cress (p.27), with similar *roundly-lobed leaves* in a basal rosette, and narrower-lobed stem leaves, but its flowers are *lilac, pale pink or white* and up to 2 cm across, with *petals 2–3 times as long as the sepals*.

Amphibious Bistort
Persicaria amphibia
KNOTWEED FAMILY
Fl: 7-9 Ht: 30-75 cm

This species grows in or beside ponds, canals and slow-flowing rivers across Scotland, but is uncommon in the C and NW Highlands. The aquatic form has *floating stems* and *floating oblong leaves*, abruptly cut off at the base. The terrestrial form has upright stems and *leaves* with a *rounded base*. Both have *slender spikes*, 2–4 cm long, of *small pale-pink flowers*.

Water Pepper
Persicaria hydropiper
KNOTWEED FAMILY
Fl: 7-9 Ht: 25-75 cm

At the base of each *narrow lance-shaped leaf*, 3–9 cm long, the stem of Water Pepper is enwrapped by a *papery, brown sheath*. The *slender, nodding inflorescences* have many flowers with 5 greenish lobes, *covered in yellowish glands*. The plant grows round ponds and lochs and in damp places over most of Scotland except the far N, and is rare on the Isles.

Ragged Robin
Silene flos-cuculi
PINK FAMILY
Fl: 5-6 Ht: 30-75 cm

Found in damp grassland, marshes and ditches throughout Scotland, except for Highland areas, Ragged Robin has *opposite, unstalked, lance-shaped leaves*. Its campion-like flowers have a slightly inflated calyx tube, and 5 *rosy-pink (or occasionally white) petals*. Each petal is *finely divided into four thin segments*, giving the flowers a tattered, lacy appearance.

Blinks
Montia fontana
BLINKS FAMILY
Fl: 5-10 Ht: 2-50 cm

This inconspicuous plant forms *compact tufts* on dry land and *straggling, rooting or floating stems* in water. It has *spatula-shaped* leaves and *tiny white flowers, 3 mm across*, clustered at stem tips, with 2 sepals and 5 *unequal petals*. These ripen into *tiny, round fruit capsules*. It grows in springs, marshes, wet grassland and cut peat surfaces throughout Scotland.

Indian Balsam
Impatiens glandulifera
BALSAM FAMILY
Fl: 7-10 Ht: 1-2 m

This robust annual with sturdy *reddish stems* was introduced from the Himalayas in 1839, reaching Scotland by 1920. It forms patches on riverbanks and wasteground N to Shetland but with few sites in the NW and none on the W Isles. It has *toothed, lance-shaped leaves* and *long-stalked, hooded, purplish-pink flowers*, shaped like Chinese lanterns, to 4 cm long.

Yellow Loosestrife
Lysimachia vulgaris
PRIMROSE FAMILY

Fl: 7-8 Ht: 60-150 cm

This robust perennial forms patches on riverbanks, loch shores and marshland, scattered around Scotland except for the Isles. Its *downy leaves, in whorls of 2–4,* are *lance-shaped, almost stalkless,* 3–10 cm *long,* and often *black- or orange-dotted.* Its *flowers, in a narrow spike,* are about 15 mm across, with a *5-lobed petal-tube* and *5 orange-rimmed sepal-teeth.*

Brookweed
Samolus valerandi
PRIMROSE FAMILY

Fl: 6-8 Ht: 5-45 cm

Confined to streams, ditches and marshes near W coasts, from the Solway to the W Isles, and E coasts S of the Forth, this *pale-green perennial* has *spoon-shaped leaves,* to 8 cm long, in a *basal rosette* and *spiralling* up the stem. Its *long-stalked, white* flowers, in a *branching spike,* are about 3 mm across, with *5 petals joined half-way into a cup-shaped tube.*

Marsh Bedstraw
Galium palustre
BEDSTRAW FAMILY
Fl: 6-7 Ht: 15-120 cm

This *straggling, hairless* perennial has slender, *rough-angled stems*. Its *blunt-ended, lance-shaped, prickly-margined leaves*, to 1 cm or longer, grow in whorls of 4–6 and *blacken when dried*. It has *spreading, branched clusters* of white bedstraw flowers (p. 40), 2.5–5 mm across. It grows in marshes, fens (rich peatlands) and wet meadows throughout Scotland.

Common Comfrey
Symphytum officinale
BORAGE FAMILY
Fl: 5-6 Ht: 30-120 cm

The commonest of three Scottish comfreys, this grows by rivers and streams in the lowlands. Its *oval leaves*, to 25 cm long, are *stalkless*, and *their bases run down the stem as wide wings*. Its *purplish or creamy flowers*, in a *nodding, spirally-curved inflorescence*, have a *funnel-shaped, 5-lobed petal-tube*, 12–18 mm long, and *sepal-teeth twice as long as their tube*.

Water Forget-me-not

Myosotis scorpioides

BORAGE FAMILY

Fl: 5-9 Ht: 15-45 cm

The commonest of four forget-me-nots found by streams and ponds around Scotland, this species has stems with *some spreading hairs and some appressed against the stem*. It has *downy, oval leaves*, to 10 cm long, and typical forget-me-not inflorescences (p. 42), with a sky-blue petal-tube *4–8 mm across*, and a *sepal-tube with broad triangular teeth*.

Marsh Speedwell

Veronica scutellata

SPEEDWELL FAMILY

Fl: 6-8 Ht: 10-50 cm

This straggling plant inhabits ponds, bogs and wet grassland scattered throughout Scotland. It has *narrow, stalkless leaves*, to 4 cm long, *slightly clasping the stem* and often purple-tinged. In the angles of upper leaves it produces a *single inflorescence* with *up to 10 white, pale blue or pinkish flowers*. Three other speedwells grow in wet places (and see overleaf).

Brooklime
Veronica beccabunga
SPEEDWELL FAMILY
Fl: 5-9 Ht: 20-60 cm

This *hairless perennial* often forms dense patches in slow-moving streams, ditches, ponds, and marshes in the S, E and the N Isles, but more rarely in the NW and W Isles. Its *stems creep and root*, then ascend, with *opposite pairs of fleshy, oval leaves. Paired inflorescences* emerge from the angles of its upper leaves, with *10–30* blue flowers, *5–7 mm across.*

Marestail
Hippuris vulgaris
MARESTAIL FAMILY
Fl: 6-7 Ht: creeping 25-150 cm

Unlike the similar Horsetails – fern allies which reproduce by spores from terminal cones – Marestail has *tiny, green, petalless flowers* with red anthers, in the angles of leaves above the water surface. Most of its stem is submerged, with *whorls of 6–12 soft, strap-like leaves.* It grows in lochs, ponds and slow-moving streams, mostly in the S and E and the Isles.

Marsh Woundwort

Stachys palustris

DEADNETTLE FAMILY

Fl: 7-9 Ht: 40-100 cm

Resembling Hedge Woundwort (p.46), but *not foul-smelling*, this species has *paler pinkish-purple flowers* and *narrower, mostly stalkless, hairy leaves*. Its flowers are *2-lipped with a blotched lower lip*, arranged in *whorls of 6 flowers* in a dense spike with a *few, more distant whorls* below. It grows in marshes, ditches and streamsides throughout Scotland.

Skullcap

Scutellaria galericulata

DEADNETTLE FAMILY

Fl: 6-9 Ht: 15-50 cm

This *downy* perennial grows in streamsides, fens and wet meadows in the W and W islands, more scattered elsewhere and not in Shetland. It has *short-stalked, spear-shaped leaves, 2–5 cm long*, and pairs of flowers which *hang to one side from leafy bracts* in the upper stem. They have a *2-lipped, blue-violet petal-tube*, much longer than the 2-lipped sepal-tube.

MINTS

Mints (*Mentha* species), such as Corn Mint (right), have a petal-tube with 4 nearly-equal lobes and a sepal-tube with 4–5 equal teeth. The small, bell-shaped flowers grow in dense inflorescences. Aromatic when fresh, the plants spread by runners, forming patches in wet places. They hybridise freely, and the hybrids are often cultivated for culinary use.

Corn Mint (*M. arvensis*) (illustrated), with *sickly-smelling, hairy leaves* and *lilac flowers* in the *angles of lower leaves*, grows uncommonly in damp arable fields scattered around Scotland.

Water Mint
Mentha aquatica

DEADNETTLE FAMILY

Fl: 7-10 Ht: 15-90 cm

The commonest Scottish mint, this *hairy* perennial grows in ponds, ditches, streams, marshes and wet woods around Scotland, except for parts of the Highlands. It has *purple stems* with pairs of *mint-scented, oval leaves*, 2–6 cm long. Its *lilac-pink flowers*, with *long stamens* and a *downy sepal-tube* with 5 triangular teeth, grow in a *rounded head*.

Spear Mint
Mentha spicata
DEADNETTLE FAMILY
Fl: 8-9 Ht: 30-90 cm

Found in damp disturbed ground through the S, E and far W, this *spearmint-smelling plant* has *hairless stems* with *opposite pairs of slightly hairy, lance-shaped leaves*. Its *pink or white flowers* have *narrow, hairy sepal-teeth to 3 mm long* and *long stamens*. Its stems are topped by *interrupted* flower clusters. Several other species and hybrids also grow in Scotland.

Gipsywort
Lycopus europaeus
DEADNETTLE FAMILY
Fl: 6-9 Ht: 30-100 cm

This *slightly hairy*, mint-like perennial has *unstalked, lance-shaped leaves,* to 10 cm long, *with narrow, forward-directed lobes*. Its *flowers, clustered at the leaf bases*, have a *5-toothed sepal-tube* almost enclosing a *white petal-tube* with *4, similar-sized, purple-dotted lobes*. It grows in marshes and watersides in the W and near E coasts.

Monkey-flower
Mimulus guttatus
MONKEY-FLOWER FAMILY

Fl: 7-9 Ht: 20-50 cm

The commonest of several *Mimulus* garden escapes from N America, known in the wild since 1824, this grows in streambanks throughout Scotland, especially in the E. It has *sprawling stems*, with *broad, irregularly-toothed leaves*. Its *showy yellow flowers* have a *long petal-tube and 2 prominent lips*. The upper lip is 2-lobed and the lower 3-lobed and *red-dotted*.

Marsh Cudweed
Gnaphalium uliginosum
DAISY FAMILY

Fl: 7-8 Ht: 4-20 cm

This *low, bushy annual* has *sprawling, much-branched stems*, topped by a *dense cluster of 3–10 rayless daisy heads*, 3–4 mm long and shorter than the surrounding leaves, which are *narrow, grey and cottony*. The heads have a *central tuft of yellow florets*, surrounded by a *cup of bracts*. It grows in sporadically wet, open ground around Scotland and the Isles.

Tansy

Tanacetum vulgare

DAISY FAMILY

Fl: 7-9 Ht: 30-100 cm

Tansy is an *aromatic perennial* with *purplish stems* and *leaves* to 25 cm long, *finely divided into toothed lobes*, with the *uppermost leaves stalkless*. It produces *dense, flat-topped inflorescences* of *button-like, rayless, golden-yellow flowerheads*, 7–12 mm across. It grows on roadsides, hedgerows and wasteground, scattered around Scotland, including the Isles.

Marsh Ragwort

Senecio aquaticus

DAISY FAMILY

Fl: 7-8 Ht: 25-80 cm

The only ragwort of marshes and wet grassland in Scotland, Marsh Ragwort is widespread except in the C Highlands. It is shorter (to 80 cm) and less stiff than Common Ragwort (p.55), and has *less divided leaves, often purplish below*, with a *large end lobe*. Its inflorescence is *more open and wide-branching*, with larger golden-yellow daisy flowers, *25–40 mm across*.

Butterbur
Petasites hybridus
DAISY FAMILY
Fl: 3-5 Ht: 10-40 cm

The only native butterbur (cf p. 145), this perennial forms dense patches of *long-stalked, heart-shaped, rhubarb-like leaves*, up to 90 cm across and *greyish underneath*. Before the leaves develop, it produces *cone-shaped inflorescences* of many *tufted, rayless, pale reddish-violet* flowerheads. It grows by rivers and streams around the lowlands, W Isles and Orkney.

Marsh Valerian
Valeriana dioica
VALERIAN FAMILY
Fl: 6-8 Ht: 20-150 cm

Smaller than Common Valerian (p. 97), this slender marsh plant is confined to the SE. Its *creeping runners* produce *long-stalked, untoothed, oval root leaves* and *unstalked, deeply-lobed stem leaves*. Its stems are topped by a *branching inflorescence of funnel-shaped pink flowers*, with male flowers on separate plants from females, which ripen to a *round nutlet*.

Wild Teasel
Dipsacus fullonum
TEASEL FAMILY

Fl: 7-8 Ht: 50-200 cm

Teasel grows in wasteground and riverbanks around the lowlands. A robust, *prickly biennial*, it has *undivided, oblong basal leaves*, and pairs of *narrower, wavy-edged stem-leaves*. Its *egg-shaped flower head* is overtopped *by long, curved, prickly bracts*. The inflorescences are covered in *prickle-tipped flower bracts*, with *purplish petal-tubes* half-hidden amongst them.

Marsh Pennywort
Hydrocotyle vulgaris
PENNYWORT FAMILY

Fl: 6-8 Ht: up to 25 cm

Marsh Pennywort spreads by *slender, creeping, rooting stems*. These produce *wavy-edged, circular leaves*, 8–35 mm across, on *stalks* which *attach centrally on the leaf's underside*. The *tiny clustered flowers*, 1 mm across, have *5 green or pinkish petals*, and develop into *brown-dotted, pumpkin-shaped fruits*. It grows in bogs and marshes throughout Scotland.

Hemlock Water-dropwort

Oenanthe crocata

CARROT FAMILY

Fl: 6-7 Ht: 50-150 cm

This *stout*, *parsley-smelling* perennial resembles Hemlock (below) – and is equally poisonous – but it *grows in or by water* in the E and W lowlands, but not Shetland. It has ridged, *green stems* and *leaves divided into wedge-shaped lobes* which *slightly sheath the stem*. Its umbels (see p. 38) have *12–40 rays* and *about 5 bracts*. Its *cylindrical fruits have two long 'horns'*.

Hemlock

Conium maculatum

CARROT FAMILY

Fl: 6-7 Ht: up to 2 m

Poisonous in all parts, Hemlock has grooved, *purple-spotted stems*, *mousy-smelling foliage* when crushed, and *round fruits*, about 3 mm long, with *wavy ridges*. It has *fern-like leaves* and *spreading inflorescences*, to 5 cm across, with *10–20 rays*. It grows in damp places, hedgebanks and rubbish tips in the E, SW and W islands, but is rare in the N and N Isles.

Lesser Marshwort
Apium inundatum
CARROT FAMILY
Fl: 6-8 Ht: 10-50 cm

This *straggling or floating perennial* grows partly or completely submerged in lochs, ponds and ditches, scattered through the lowlands and Isles. Its *submerged leaves are divided into thread-like segments* and its *aerial leaves have about seven 3-lobed leaflets*. The white-flowered *inflorescences branch off the stem opposite a leaf* in summer, with *2–4 rays and no bracts*.

Cowbane
Cicuta virosa
CARROT FAMILY
Fl: 7-8 Ht: 30-130 cm

This poisonous perennial grows uncommonly in shallow water and lochsides in S and C Scotland and on S Uist. It has *ridged, hairless, hollow stems*. Its leaves on *long stalks* are divided into *sharply-toothed, spear-shaped segments*. Its *inflorescences* are *7–13 cm across*, with *10–30 rays*, and its *globular, ridged fruits* have a *collar of sepal-teeth* and *2 long styles*.

Giant Hogweed
Heracleum mantegazzianum
CARROT FAMILY
Fl: 6-8 Ht: up to 3.5 m

Introduced to Victorian gardens, this massive alien today grows, often abundantly, by rivers and in wasteground in the SW, in the E as far N as Caithness, and on Shetland. It has *hollow, purple-spotted stems* reaching *10 cm across, lobed leaves to 2.5 m long,* inflorescences up to *50 cm across,* with *white or pinkish flowers,* and *oval, winged fruits about 1 cm long.*

Common Duckweed
Lemna minor
DUCKWEED FAMILY
Fl: 6-7; Ht: floating

This tiny, floating waterplant sometimes forms green sheets over ponds, ditches and canals, mostly in the S and more scattered in the NE, W islands and N Isles. A *single root,* to 15 cm long, dangles below each *round, 3-veined frond, 2–5 mm long.* On top of the fronds it irregularly produces *minute greenish flowers,* but it reproduces mainly by budding off new fronds.

Water-plantain
Alisma plantago-aquatica
WATER-PLANTAIN FAMILY
Fl: 6-8 Ht: 20-100 cm

This aquatic perennial has a *basal tuft of long-stalked, broadly oval, plantain-like leaves and leafless flowering stems*, topped in summer by *several whorls of long, branching flowerstalks*. Its *flowers are 7–12 mm across*, with *3 pale-lilac petals* and 3 sepals. It grows on mud in shallow water in S and C Scotland and near the Moray Firth.

Canadian Waterweed
Elodea canadensis
FROGBIT FAMILY
Fl: 5-10 Ht: up to 3 m underwater

Also known as Canadian Pondweed, this *submerged aquatic*, discarded from fishtanks, has become common in slow-moving freshwater in the lowlands and Orkney. It has *brittle stems*, with *whorls of 3 oblong, translucent leaves*, 5–12 mm long. It infrequently produces *floating female flowers with 3 narrow white or pale-purple petals*. Male flowers are rare.

Bog Pondweed
Potamogeton polygonifolius
PONDWEED FAMILY
Fl: 5-10 Ht: up to 50 cm underwater

At least 18 species of Pondweeds grow in Scotland, but this is by far the commonest, in bog-pools, ditches and ponds throughout the country. All produce rather knobbly *spikes*, held above the water in summer, of *tiny greenish flowers* with 4 sepals, *no petals* and 4 stamens. This species has *reddish, oval floating leaves* and *spear-shaped submerged leaves*.

Yellow Iris
Iris pseudacorus
IRIS FAMILY
Fl: 5-7 Ht: 40-150 cm

Yellow (Flag) Iris is common in wet ground except in mountainous areas. It forms dense patches of *sword-shaped, bluish-green leaves*, 40 cm or more long. Its *stout stems* are topped by a branched inflorescence of *4–12 yellow flowers* to 10 cm across, with leaf-like bracts. The complex flowers have 6 *flower-lobes* and *3 style branches that look petal-like*.

Branched Bur-reed
Sparganium erectum
BULRUSH FAMILY
Fl: 6-8 Ht: 30-150 cm

This *upright* perennial of still and slow-moving water and marshland in the S, E, W islands and N Isles, has *keeled, strap-shaped, sheathing leaves*, to 15 mm wide. Its *branched flower-spikes* have *unstalked, rounded heads of tiny, 3–6-lobed flowers*. The *smaller, upper heads are male* with *yellow stamens*. The *broader female heads* below ripen into bur-like fruits.

Bulrush
Typha latifolia
BULRUSH FAMILY
Fl: 6-7 Ht: 1.5-2.5 m

Although not the Bulrush of the Bible, the name has stuck to this robust perennial – also called Reedmace – of reed-swamps, lochs and riversides in the S and E. Its *strap-shaped, greyish-green leaves* overtop an *unbranched flower-stem* with a *cylindrical, chocolate-brown spike* of female flowers, immediately beneath a narrower, *fluffy, straw-coloured male spike*.

PLANTS OF MOORLANDS AND UPLANDS

The sweeping moorlands, sustained by the cool and often wet climate and thin soils, are one of the glories of Scotland's countryside, providing a colourful backdrop to the mountains beyond. Although many unusual and attractive species grow here, they are rarely abundant. There is only space in this *Mini Guide* to cover the commoner and more characteristic species that a visitor to this harsh landscape is likely to meet.

Many parts of the Highlands are at their most colourful in late August and early September when Heather is in full bloom, transforming whole hillsides into glorious sheets of purple (see p. 9). But this spectacle is largely man-made. By nature, Heather is an understorey shrub of pine woodland, but almost all the native pinewoods of Scotland have long since been felled. Open heather moors are maintained by burning (see photo above), to provide food and habitat for the Red Grouse that are shot for sport on Highland estates.

Patches of purple on the hill earlier in the year are more likely to belong to Bell Heather, which is distinguished from Heather by its larger, richer-red flowers. It is a plant of drier peaty soils, replaced in boggier moorland by Cross-leaved Heath. Other members of the Heather family also grow on these moors. Bilberry (Blaeberry in Scots) prefers drier moors, where its deciduous leaves are often much-grazed by sheep. On slightly damper and more exposed moors, it is joined or replaced by the evergreen Cowberry. In the north-east, Bearberry – another heath with evergreen leaves – is also locally abundant. Various kinds of purple orchids add further colour to this landscape.

The Flow Country in Caithness and Sutherland is the largest and most important peatland area in the UK and home to many moorland plants.

*A few 'Old Caledonian' woods of Scots Pine (*Pinus sylvestris*) still survive, as here below Beinn Eighe in Wester Ross.*

Where some native Scots Pine woodland does survive – most notably around Deeside and Speyside – characteristic pinewood flowers also occur. These include four species of Wintergreen and the unrelated Chickweed Wintergreen, which also survive out onto open moorland. Several are now very localised and scarce, such as the delicate, creeping Twinflower (*Linnaea borealis*).

The high rainfall of the west Highlands (reaching 198 cm in some areas) may seem inconvenient to visitors, but it is also an important part of the Highland scene, shaping and colouring the landscape. Where the rainfall exceeds about 140 cm (55 ins), bog mosses (*Sphagnum* species) become dominant and, over the centuries, have formed vast blankets of peat. In these bogs, Cotton-grasses, Sundews, Butterwort, Lousewort and the shrubby Bog Myrtle are amongst the commoner flowering plants.

Interspersed with the moors, woods and bogs, many lochs add reflected glory to the Highland landscape. Their cold, acid waters generally support few plants, but, in some, the leaves and flowers of White or Yellow Water-lilies may be seen floating on the surface. In early summer some shallower lochs, pools and bogs are pink with the frilled flowers of Bogbean.

*One-flowered Wintergreen (*Moneses uniflora*) (left) and Twinflower (*Linnaea borealis*) (right) are two scarce species from the understorey of pine woods.*

Heath Milkwort
Polygala serpyllifolia
MILKWORT FAMILY
Fl: 5-8 Ht: 10–30 cm

This slender perennial grows on heaths and poor grassland throughout Scotland. It has *unstalked, narrowly oval leaves, the lowest of which are opposite* (although shed by late season). Its spikes of *3–10 blue, pink or white flowers* are 4–6 mm long. They have *3 fused petals*, and 5 sepals, the inner 2 of which are *larger, petal-coloured and enclose the flower*.

Bog Myrtle
Myrica gale
BOG MYRTLE FAMILY
Fl: 4-6 Ht: 60–150 cm

Willow-like in appearance, this patch-forming shrub of bogs and wet heathland throughout the SW, Highlands and islands, except Shetland, is recognisable by the *resinous smell* of its crushed leaves. Its *catkins* appear on leafless twigs in spring. Its oval leaves, which appear later, are *greyish-green, toothed near the tip*, and *dotted with yellow glands underneath*.

Marsh Violet
Viola palustris
VIOLET FAMILY

Fl: 4-7 Ht: 2–4 cm

This *hairless* violet produces a tuft of *almost circular, wavy-edged leaves, to 4 cm long*, from a creeping underground stem. Its flowers, on *stalks no longer than the leaves*, are 10–15 mm vertically, with *pale-lilac* (rarely white), *dark-veined petals* and a *spur little longer than the backward-pointing sepal flaps*. It grows in bogs and marshy ground throughout Scotland.

Slender St John's–wort
Hypericum pulchrum
ST JOHN'S-WORT FAMILY

Fl: 6-8 Ht: 30–60 cm

This delicate perennial grows in heathy and grassy places on acid soils across Scotland. It has *slender, reddish stems, without ridges or wings* (cf species on p.75), and *paired, oval leaves*, 5–20 mm long, which *half-clasp the stem* and are *dotted with translucent glands*. Its flowers, about 15 mm across, have 5 *reddish-yellow petals, dotted with black marginal glands*.

Round-leaved Sundew
Drosera rotundifolia
SUNDEW FAMILY
Fl: 6-8 Ht: 6–25 cm

The commonest of three *carnivorous* Sundews, all with *reddish leaf hairs* which exude drops of sweet, sticky juice to trap insects, this grows in wet peaty places around Scotland. Its *rounded leaves, 4–10 mm across*, narrow into *a hairy stalk*. Its flowers, on leafless stalks to 10 cm tall, are about 6 mm across, with 5 or 6 white petals, but often remain tightly closed.

Chickweed Wintergreen
Trientalis europaea
PRIMROSE FAMILY
Fl: 6-7 Ht: 10–25 cm

A *whorl of 5–6 shining, oval leaves,* 1–8 cm long, grow *near the top* of the slender stems of this delicate perennial, with a few, small, alternating leaves lower down. Its flowers are *usually solitary*, 15–18 mm across, with *5 to 9 (typically 7) white petal-lobes*. It grows in pinewoods and open, mossy moors scattered through the Highlands, N Isles and S hills.

Bearberry

Arctostaphylos uva-ursi

HEATHER FAMILY

Fl: 7-9 Ht: creeping to 1.5 m

Bearberry is a *mat-forming shrub* with *evergreen, untoothed, oval leaves*. These are *dark-green and glossy above* and *paler and veined beneath*. Its flowers are *urn-shaped*, 4–6 mm long, and *pale pink* with a *deep rose rim*. They ripen into *glossy red berries*. It grows on dry moors scattered across the Highlands and Isles, but is rare in the Border and Galloway hills.

Crowberry

Empetrum nigrum

HEATHER FAMILY

Fl: 5-7 Ht: 15–45 cm

This *mat-forming shrub* grows on peaty and rocky moorland, mountain plateaux and clifftop heaths throughout Scotland. Its stems are densely covered in *short-stalked, hairless, needle-shaped leaves*, 4–7 mm long. Its *easily-overlooked flowers*, 1–2 mm across with 6 *purplish lobes*, grow in the angles of stem-tip leaves, and ripen into a *round, black, bitter berry*.

Cross-leaved Heath

Erica tetralix

HEATHER FAMILY

Fl: 7-9 Ht: up to 60 cm

This *evergreen shrub* prefers wetter ground than Bell Heather (below) in bogs and moorland throughout Scotland, although both grow together where soil dampness is variable. It has *greyish, hairy, needle-shaped leaves*, in *cross-like whorls of 4* up the stem, and *compact, 1-sided clusters of drooping flowers*, 6–7 mm long, with an *inflated, pale pink petal-tube*.

Bell Heather

Erica cinerea

HEATHER FAMILY

Fl: 7-9 Ht: up to 60 cm

Often abundant on drier moorland throughout Scotland, Bell Heather has *larger flowers* than Cross-leaved Heath (above), in whorled spikes, with a *deep reddish-purple, urn-shaped petal-tube*, 5–6 mm long, and 4 short, purple sepal-lobes. They develop into *many-seeded dry capsules*. Its *needle-shaped leaves* grow in *whorls of 3* or *bunched into knot-like clusters*.

Heather
Calluna vulgaris
HEATHER FAMILY
Fl: 7-9 Ht: up to 60 cm

This bushy shrub is abundant on open moorland, thanks to grouse management (see p. 188). Its *leafless stems* have *many short side shoots, with dense, opposite rows of needle-shaped leaves* to 2 mm long. Its flowers, in *crowded, leafy spikes* (see p. 9), have *4 pinkish-purple sepal-lobes*, and a *short, pink, bell-shaped petal-tube*. They develop into *round, dry capsules*.

Cowberry
Vaccinium vitis-idaea
HEATHER FAMILY
Fl: 6-8 Ht: up to 30 cm

Cowberry is common in moors and woods on acid soils in the Highlands and S Uplands, and scattered on the islands. It has *dark-green, leathery, evergreen leaves*. These resemble Bearberry leaves (p. 195), but are *broadest in the middle* with *dark glands dotting the underside*. Its nodding groups of *pinkish-white, bell-shaped flowers* ripen into *red berries*.

Blaeberry
Vaccinium myrtillus

HEATHER FAMILY

Fl: 5-7 Ht: up to 60 cm

Known as Bilberry in England, Blaeberry grows on moors and in open woods on acid soils throughout Scotland. It has *angled, green stems* and *thin, deciduous, bright-green, toothed, oval leaves*, 1–3 cm long. Its single or paired, *nodding, globe-shaped flowers*, to 6 mm long, are *greenish-red*, with *short, curled-back lobes*. They develop into *edible bluish-black berries*.

Common Wintergreen
Pyrola minor

HEATHER FAMILY

Fl: 6-8 Ht: 10–30cm

Wintergreens are three species of *evergreen* perennials (hence their name) which grow in pine and birch woods, open moorland and coastal dunes scattered around Scotland. This species has *slightly-toothed, broadly oval leaves* with *stalks shorter than their blades*. Its *globe-shaped, white or pinkish* flowers, are *about 6 mm across* and borne in dense spikes.

Heath Bedstraw
Galium saxatile
BEDSTRAW FAMILY
Fl: 6-8 Ht: 10–20 cm

This *slender, scrambling or mat-forming* bedstraw is abundant in heathland, woods and grassland on acid soils throughout Scotland. It has *smooth, 4-angled stems* and *whorls of 6–8 bluntly oval leaves, 7–10 mm* long. Its white flowers, 3 mm across, are clustered on paired stalks in the angles of leaves. A similar species ifs found, less commonly, on limestone.

Wood Sage
Teucrium scorodonia
THYME FAMILY
Fl: 7-9 Ht: 15–30 cm

Wood Sage grows in dry, shady heathland and rocky places on lime-free soils throughout Scotland, apart from Shetland and parts of the NE. Its stems, to 30 cm tall, have opposite pairs of *downy, wrinkled, heart-shaped leaves*, topped by slender spikes of *greenish-yellow flowers, 8–9 mm* long, in the angles of short leafy bracts, with *4 protruding scarlet stamens*.

Lousewort
Pedicularis sylvatica
BROOMRAPE FAMILY
Fl: 4-8 Ht: 8–25 cm

A plant of damp heaths, bogs and marshes across Scotland, Lousewort is *8–25 cm tall* with many *spreading branches*. Its purplish, *oblong* leaves are divided into many *narrow, deeply-toothed lobes*. It has *spikes of 3–10 flowers* with a *pink, red or rarely white, 2-lipped petal-tube, 20–25 mm long*. The taller, more erect Red-rattle (*P. palustris*) grows in wetter bogs.

Common Butterwort
Pinguicula vulgaris
BLADDERWORT FAMILY
Fl: 5-8 Ht: 5–18 cm

This *carnivorous* plant grows in bogs, wet heaths and on bare wet rocks around Scotland. Its *oblong, yellow-green leaves, 2–8 cm long*, are covered in glistening *sticky glands* which trap and digest small insects. Its leafless flowerstalks are topped by a *solitary purple flower*, with a *3-lobed* lower petal-lip, *paler near the throat*, and upper lip with *2 backward-curved lobes*.

Bog Asphodel
Narthecium ossifragum
BOG ASPHODEL FAMILY

Fl: 7-9 Ht: 5–40 cm

Patches of Bog Asphodel grow in bogs and wet moors throughout Scotland, except parts of the SE. Its flowering stem, with a few short, sheathing leaves, grows from a tuft of *rigid, strap-like basal leaves*. It has a *dense spike* of 6–20 flowers, to 15 mm across, with *2 whorls of 3 narrow, yellow, petal-like lobes*. These ripen to *orange, spindle-shaped fruit capsules*.

Common Cottongrass
Eriophorum angustifolium
SEDGE FAMILY

Fl: 5-7 Ht: 20–60 cm

This sedge is conspicuous in bogs and bog pools throughout Scotland. Its grass-like leaves turn red from the tip. Its flower stems end in a cluster of 3–7 *dropping heads* of brownish-green flowers which develop into *shining white balls* of long, cottony hairs on the fruits. Harestail Cottongrass (*E. vaginatum*) with *one, egg-shaped* head is common in blanket bogs.

Heath Spotted Orchid
Dactylorhiza maculata
ORCHID FAMILY
Fl: 6-8 Ht: 15–60 cm

This common Scottish orchid grows on acid moorland, marshes and open woodland. It has up to 8 *strap-shaped leaves, marked with pale purple blotches*, and a pyramidal head of 5–20 flowers. These are *white to pale lilac* with *deeper pink dots and lines on the lower lip. The centre lobe of the lower lip is much smaller than the rounded side lobes* (cf species on p.101).

Early Marsh Orchid
Dactylorhiza incarnata
ORCHID FAMILY
Fl: 5-7 Ht: 15-50 cm

This variable orchid has 3–7 strap-shaped, *yellowish-green, keeled leaves*, and inflorescences of 10–40 flowers, coloured from *flesh pink to deep magenta*. These have a *shallowly 3-lobed lower lip, bent back at the sides* and marked with *loops and dots*. It grows in marshes and damp grassland scattered across Scotland. Several other purple orchid species also occur.

White Waterlily
Nymphaea alba
WATERLILY FAMILY
Fl: 7-8 Water depth: 0.5–3 m

This showy perennial grows in nutrient-poor lochs and ponds around Scotland except Orkney. Its *round, floating leaves* are 10–30 cm across. The *floating flowers* are 10–20 cm across, with 3–5 sepals, which are white above and olive-green beneath, and *20–25 large, white or pink-tinged petals*. Yellow Waterlily (*Nuphar lutea*) is found, less commonly, in the S.

Bogbean
Menyanthes trifoliata
BOGBEAN FAMILY
Fl: 5-7 Ht: 12–30cm

This plant forms floating mats in ponds and shallow lochs, or creeps through bogs, most commonly in the W. Its leaves, raised above the water on long stalks, have *3 oval, untoothed leaflets*. Its branching spikes have 10–20 *funnel-shaped flowers*, 15–20 mm across, with *5 pink petal-lobes and a fringe of white hairs at the throat*, ripening to a *bean-shaped fruit*.

Index
Common Names
Page references in italics refer to photographs